A Convergence of Two Minds

Origins of Self-Awareness and Identity

A Convergence of Two Minds

Origins of Self-Awareness and Identity

Randolph R. Croxton

PALUSTRIS

For information about permission to reproduce selections from this book, please write to
Palustris Press, 475 Fifth Avenue, New York, NY 10017
email: contact@palustrispress.com

I I I I I I I I

Cover and Book design by Jean Hahn

Cover image: Hand of God (Auguste Rodin, detail of bronze casting)
Carnegie Museum, Pittsburgh, PA
Copyright © 2015 Carnegie Museum

Figure 2.5: Photograph, Hall of Bulls, Lascaux, France
Copyright © 2015 Sisse Brimberg/ National Geographic.

I I I I I I I I

Library of Congress Control Number: 2015911260

ISBN 978-0-9961176-0-9

First Edition

I I I I I I I I

Palustris Press
475 Fifth Avenue, New York, N.Y. 10017
www.palustrispress.com

*Dedicated to my wife, Fran Drummond
whose combination of intelligence and grace
makes all things possible.*

Contents

Foreword

HAVE YOU EVER WONDERED IF YOUR PATHWAY IN LIFE IS THE ONE you were meant to follow? In 1954 my fourth grade teacher, Mrs. Marjorie Brown, pulled me aside in class to say that since I was good at art and math—a rare combination—"Maybe you should be an architect." I have often wondered how she could possibly have foretold my career at age ten and why my sister, my cousins, and now my two sons have no interest in math. Observing that each one of us, even brother and sister, is born with a unique set of talents and predispositions has led to a second field of interest: the natural sciences and, of special interest, evolution and human behavior.

Fortunately, as an architect, I have been able to join these passions in an approach to design based on humanistic and ecological function: environments for work, learning, and healing. Needing to understand both the built and natural dimensions of this approach has driven me to read deeply into the natural sciences—the life sciences and the closely related social sciences—which consider Darwinian evolution and natural selection to be the basis for understanding human nature, mind, and body.

However, Darwin's exquisite explanatory model, for all its utility as a scientific framework, does not seem to be tracking what we observe in life. Rather than the expected pattern of families with many examples of mathematical genius, or brilliant musicianship, or artistic prowess, these qualities seem to be occurring in a more random manner. Families display wide variation, even in the extreme example of identical twins—with identical DNA—where there is consistent evidence of individual differences in mind, if not in body. There appears to be an unrecognized effect, beyond the random mutations or environmental effects that are often used to explain away such differences.

Organization by specialty and breaking down complex natural processes has been a hallmark of success in science, whereas architecture is a holistic and deeply integrative process. Over the years and from this point of view, I have observed a pattern of avoidance in critical areas of study within the natural sciences, which, because of their controversial or "third rail" nature, are underfunded or not supported by government or scientific institutions. The all-too-human characteristics of warfare, interpersonal violence, racism, male dominance and associated abuse/kidnapping/rape of women, not to mention homosexuality, sexual preference, and gender identity, have been characterized as deviations from the norm. These are further examples of the "differentness" among humans that have so intrigued me.

A few courageous individuals in the life sciences have attributed this variability to an evolutionary process. However, the full range of these features, their evolutionary value, their biological pathways, and their expression in patterns of "differentness" maintained across the generations has gone unrecognized. These areas of avoidance, I believe, contain bridges to a deeper understanding—the recognition that they are cause-and-effect elements in our emergence as the one fully self-aware species on Earth.

The first barrier to understanding the rise of humankind is the belief that we possess a single "human nature." We are the product of sexual reproduction, and our external reality—our life experience and how the world relates to us—is set by our anatomical gender. The nature of that experience is one of two very different realities: male or female. By contrast, our internal reality—how we perceive the world—flows from the complex legacy of both male and female ancestors: the nature part of "nature versus nurture." Both of these dimensions of gender inform who we are, therefore there are an infinite number of possible mindsets or "personalities," which we see every day at home, at work, and on the evening news—from saints to psychopaths.

As simple as this may sound, it is a significant step beyond our fixation on what is "normal," and it is a precondition to understanding our evolution and the source of our respective natures. That having been said, we are not slaves to our natures, but exercise a degree of free will that is an integral part of our self-aware mind: our being conscious of, and learning from, our surrounding environment. *Nature* and *Nurture* are the simultaneous and often opposing sides of our internal conversation, the debate between our inherited ancestral wisdoms and the wisdom we acquire in life.

The second barrier to understanding our rise to humanity is the long fixation of science on the elegant strands of our protein coding DNA (the familiar double helix), which constitutes approximately 1% of the genetic material in the human genome (the complete set of our genetic information). Unfortunately, this narrow focus has served to reinforce the consensus that there is a "normal" brain. While the outward features of the brain and the processes of natural selection suggest that we should possess a fairly uniform and predictably complex brain, it is now coming to light that there is a wizard behind the curtain.

That wizard is the remaining 99% of the human genome (non–protein coding DNA) thought to be a relatively inactive cousin of DNA and characterized as "junk DNA" early on. As proposed here, that "junk" contains the composer of the vast spectrum of the personalities of humankind.

Carl Zimmer, in his article "Darwin's Junkyard" in the *New York Times Magazine* of March 8, 2015, outlines the continuing debate among biologists on the value and functionality of what is still called junk DNA by some. One side of the argument, captured in the article, continues to insist that it is essentially nonfunctional: "…useless copies of genes and new transposable elements." On the other side of the argument: "Because a few of these RNA molecules [in junk DNA] have turned out to be so crucial…the rest of the noncoding genome must be crammed with riches." In this endeavor, we take the position that those riches exist, and undertake an exploration of the nature, foundational role, and existential value of those riches to humankind.

Our brain is divided into two, often opposing, worldviews. Each half is an inherited "DNA sourcebook," and, taken together, they contain the full wisdom of our evolutionary heritage. For this reason, our brains look similar at first glance, as we would have expected from a DNA/natural selection process. However, they are interconnected at levels of complexity and subtle interpretation that, to date, have not been fully understood. The primary author of these sourcebooks, as made clear on the following pages, is gender—the life experiences of our male and female ancestors—passed down to us by natural selection/DNA. The editing and joining of these sourcebooks to create a unique mindset is the province of RNA.

Our differentness has a pattern; we are not wildly and randomly different but, rather, maintain many proportional and repetitive "varieties" from generation to generation. For instance, if we look at the proportions of the widest range of variables—liberals, conservatives, homosexuals, savants, autistics, sociopaths, or psychopaths—we find that they keep occuring in the same general percentages of the population over time. This has led some scientists to observe that there must be an evolutionary or adaptive benefit to this spectrum of differences or they wouldn't keep recurring in the same proportions. The answer is that they provide a defining evolutionary advantage. The biological basis and unexpected connection to gender in this pattern of difference will be explored at the scale of the individual and at the scales of our cultural, corporate, and governmental institutions.

In evolutionary terms, our human story begins quite recently, approximately 100,000 years ago, with the dawning of self-awareness and the projection of our consciousness into the surrounding world. At that point, the pathways of survival for males and females in the previous seven million years of the hyper violent environments of our ancestry, had conserved two very different sets of skills, perceptions, and sexual preference which were captured as the two "gendered" hemispheres of our brains. However, at this crucial juncture of our emergence, there was a change; the near-total dominance of the survivalist male lineage began to rebalance with the ascending social/creative female lineage. This combination of world views was the gateway to self-awareness which empowered our elimination of all competing members of our extended human family and allowed our ancestors to possess every corner of the world.

The significance of this new construct is that there is no such thing as a "normal" brain and that our male and female ancestors are ever present in the pathways of our thought. The wide range of personalities that flow from the combination of the gendered hemispheres of our brain are essential to a resilient population, assuring through our differentness—the sexual and non-sexual dimensions of gender—that humankind is sufficiently diverse to survive in times of catastrophe and thrive in times of opportunity.

The strongly held interpretations in this book are reinforced by the integration of scientific source materials and linked together by logic, hypothesis and working assumptions. This is not a work that achieves the rigor of the scientific method; rather, it is an integrated reading of the observed world from the point of view of someone who has spent his career at the intersection of human nature and the natural world. This endeavor points to the specific areas of scientific investigation that will ultimately confirm or refute this understanding of our human natures.

The excitement of these connections and the immediate relevance of such a unifying conception—gender as humankind's fabric of perception and behavior—have compelled an architect to put common sense aside and undertake the writing of a book in the natural sciences.

Randolph R. Croxton

Introduction

EVOLUTION, ACTING THROUGH NATURAL SELECTION, IS THOUGHT OF as taking place over hundreds of thousands of years in the gradual refinement of our physical form and function. In short, we see it as occurring so slowly as to be irrelevant in our daily lives. Unrecognized until now, there is a second evolutionary force that has been enhancing our collective chances of survival and, in this case, moving at lightning speed just before and after our birth. Operating at the foundational levels of the mind, it shapes our unique view of the world. Nothing could be more immediately relevant.

There are two architects of our mind, both of which exist in the human genome (the complete set of our genetic information). DNA, the most familiar, was thought of in the 1990s as the "master script" of survival wisdom handed down in each human lineage, although it represented only 1.2 percent of the genome. The somewhat mysterious partner of DNA, given the unflattering title of "junk DNA" by some, was considered to be a relatively inconsequential wasteland of genetic remnants, even though it comprised 98.8 percent of the genome. All uncertainty on the matter was to be erased in early 2001 with the completion of the mapping of a whole human DNA sequence by the Human Genome Project, an endeavor of international scope.

THE EVENT

A milestone event in the field of evolution and biology occurred in February 2001,[1] the heralded mapping of the human genome was published, and our

DNA, the all-powerful and complex blueprint for humans, was finally made known. This was, among many other objectives, expected to be the threshold event in the elimination of hereditary cancer, diabetes, and heart disease. In one of the greatest missed calls of modern scientific endeavor, our DNA was found to be less complex than some species of insects, even of some plants!

As a simplification of what was found: DNA *(protein coding DNA)* can be thought of as the physical instruments in an orchestra that were actually being played (combined, sequentially turned on/off, and given emphasis over time) by other composers of the music in the genome *(i.e. non-protein coding DNA)* previously characterized as "junk DNA." In other words, the essentially unknown portion of the genome was found to be the conductor, actually "expressing" who we are! This finding exponentially increased the complexity of human heredity as we understood it.

This event highlighted the narrow focus of the scientific community up to that date on Darwinian natural selection and protein coding DNA as the blueprint for humans. "Random mutations," "abnormalities," and "environmental factors" were the explanations given for the extreme differences of perception and behavior in the population. As translated in the culture, these people were just not "normal."

This opening of the door on the complexity of the two-part universe within the human genome struck me as an opportunity to seek a bridge between these findings and the vastly more complex range of human mindsets and behavior we observe in others.

Edward O. Wilson, University Professor Emeritus, Harvard, has observed:

> [I]t is considered probable that non-coding changes have been of key importance in the evolution of cognition, in other words the changes that make us human.[2]

Michael Snyder, director of the Stanford Center for Genomics, has stated that his research on genome characterization and gene regulation

> demonstrates that the major difference between individuals and closely related species such as human and chimpanzee is gene regulation, not gene content.[3]

Wilson and Snyder are clearly pointing to the realm of non-protein coding DNA as an important editor and interpreter of each person's genetic heritage. In the area of our interest, the neurological and behavioral sciences, a specific version of this non-protein coding DNA—non coding RNA—and brain function is determinant.[4]

The human genome consists of over 22,000 genes, but only a couple of hundred of these relate to the development of the brain and the neurological system. The actors on this much smaller but critical stage are protein coding DNA and the non coding RNA (ncRNA). For simplification and narrative flow, the protein coding DNA portion (our ancestral blueprint) will be designated DNA and the system of interpreting that blueprint—ncRNA—will be designated RNA.

As proposed herein, RNA composes and edits an individual nature for each of us, our own way of being in the world, creating a necessary variability across humankind. DNA and RNA work together during the process of our birth with natural selection/DNA acting as the slow-moving multi-generational curator and RNA acting as the instant "interpreter" of our male and female survival lineage (the hemispheres of the brain). Because DNA is genetic, it passes down viable attributes from generation to generation, while RNA is the opposite, creating a unique identity that occurs in only one person and is not passed down.

We also witness a wide range of expression of what we now call masculinity and femininity in the human population (independent of anatomical gender) that also points to this underlying pattern of variance. Through these avenues we begin to approach the complexity of perception and behavior we see across the population, a pattern we also see reflected in the distribution of human characteristics in the brain as currently mapped in the neurological sciences.

In seeking the origins of self-awareness and identity, we will be covering a time span of 100,000 years, a relatively long period by today's standards, although short in evolutionary terms. We begin with the emergence of humans from our animalistic baseline and follow our pathway to modern times.

THE EMERGENCE OF SELF-AWARENESS

The dawning of our self-awareness marks the birth of our species as human, with the ability to see ourselves as individuals in the larger context of the universe. Until then, approximately 100,000 years ago, our species had existed for the previous 100,000 years at little more than what we would call a subsistence level. For millions of years before that, our prehuman ancestors had advanced at the slow pace of biological evolution under the hard logic of natural selection. It is nearly impossible for us to appreciate the significance of the self-awareness we now enjoy: our life as an independent entity moving within an observable and changeable world. This state of existence is limitless when compared to our former brutish and instinctual world. We were little more than a survivalist mindset, operating in the narrow seams of opportunity that existed within an immutable state of nature.

Natural selection was always operating in parallel with our wide-ranging variability during our prehuman existence, but it was only with the advent of self-awareness that we began to fully project our consciousness outward into the surrounding environment. This was the breaching of a wall that now separates us from every other life form. All life on earth has been changed by us as a result, having been extinguished, impacted, or redirected. It does not overstate the case to include the further—and pressing—changes that have resulted from our actions at the scale of the earth's operational systems of climate, oceans, atmosphere, and thermodynamic equilibrium.

Other members of our ancestral human family, in their competition for survival, had developed stone weapons and used fire. I will propose, however, that we were the only ones who entered a natural-selection feedback loop initiated by our superior weaponry—most logically, projectile weapons. Such a defining advantage in the hunt for food and in the defeat of enemies consistently proves to be the path to prosperity and the immediate expansion of those mindsets. This kind of self-reinforcing cycle, acting over many generations, led to a greater concentration of our self-awareness. Those without such an advantage were at risk.

The creative and strategic advance that is inherent in the ability to project lethal force through the air was just the beginning of a flood of innovations to follow. Thereafter, the evolutionary process, which had been almost exclusively an internal, biological one, continued to transition outward to the realm of our physical, cultural, and social environments. Empowered by our manual dexterity (our ability to shape tools, weapons, and the surrounding environment), a great leap in the speed, scale, and impact of the human presence followed.

In the modern world, virtually every biological human feature seems to have been magnified by our innovations: the eye has been projected outward 46.5 billion light-years to the edge of the observable universe and inward to the subatomic constructs of matter. We can now "hear" seismic events at remote corners of the planet, and our fist, as a weapon of combat, has been extended from knife to spear to arrow to bullet to thermonuclear device in the blink of an eye in evolutionary time.

Clearly, we have hijacked an enormous portion of the evolutionary process and accelerated it to time frames unimaginable in the old internal/biological mode. Utilizing our own values and judgments, we have sidestepped an array of evolutionary end points for ourselves and our fellow humans. We have taken on a godlike role, even deciding in a medical context what constitutes a viable human being. In this manner, we have taken control over some of evolution's most powerful tools in "selection of the fittest."

And yet, to this day, there are two evolutionary legacies of our minds that have been untouched. One of these is the wide-ranging variety of our mindsets, our diverse "personalities," the ways in which each of us uniquely sees and assesses the world. The other legacy is the "internal conversation" we experience as a debate between the two sides of our nature when grappling with a difficult decision. On one side we feel empathy and long for justice, but on the other we are bedeviled by our darker side, which we so wish to leave behind. We sense that a powerful "primitive" brain is still present, so much so that we seem to walk around as two life forms: one civilized and one animal. If we were to assign a gender to each of these versions, there is little doubt as to what goes where and why we have such difficulty in "making up" our minds! Even our cartoons portray a devil (male) on one shoulder and an angel (female) on the other, each whispering suggestions in our ears. They are not there in such form or with such purpose, but nature's well-intended whispers do come.

The discovery that our brain is not only physically divided but that each hemisphere of our brain tends toward a unique and often opposite view of the world is consistent with what we intuitively know to be true. My assertion is that this duality creates a life-sustaining debate, handed down by natural selection and informed by our ancestral male and female responses to life-threatening circumstances during their respective pathways through life.

AN EVOLUTIONARY IMPERATIVE

Survival and reproduction are the primary objectives of this decision-making process, both of which require success at the core challenge of life: competition. Whether the competition is for sex, food, or land, or for prowess among our peers as the fastest runner or the most brilliant mathematician or the best actor, in other words, other ways to get sex, food, or land, these competitions—all the way up to war, the granddaddy of them all—are the substance of the human condition and the crucible of our innovation and creative/intellectual advancement.

Competition is the mother's milk of evolution; it drives the rise of physical-neurological fitness in life forms and, as an overarching feature of *Homo sapiens*, is an inherited evolutionary imperative. As the most consistently present of all biological expressions, competition is the key to understanding the rise of humankind from the dawning of our self-awareness to the ongoing emergence of democracy's competitive superiority in the global context. The rise in the scale of our presence, from a low of approximately 10,000 birthing couples in Africa at the dawning of our self-awareness to over 7 billion people today, cannot be understood absent the competitive/biological reality that drives each step:

1. We cannot fully understand the emergence of self-awareness in humans without appreciating the central role of violent competition and deadly force in prehumans.
2. We cannot fully understand the emergence of human groups beyond the small scale of biological boundaries (bands of 5–8 individuals) without understanding the need of the self-aware mind to create supernatural narratives that, through belief, remove fear and can bind large numbers of minds to a common purpose—a defining competitive advantage.
3. We cannot fully understand the emergence of complex language and symbology without comprehending their competitive advantage in further attracting and binding minds to a common purpose.
4. We cannot fully understand the destruction of nomadic hunter-gatherer groups by settled agrarian communities without recognizing the latter's greater population densities and resulting competitive advantage of superior force.
5. We cannot fully understand democracy without appreciating its elevation of the individual to self-determination, the ultimate binding force and competitive advantage over elite rule.

All of these increases in density and complexity of human organization were projections by our self-aware mind of biological features present in the prehuman mind. In this manner, the rise of humankind has been an unbroken, if uneven, ascendance toward the most efficient of all social and competitive orders, the superorganism, which has been defined as "a collection of agents which can act in concert to produce phenomena governed by the collective." [5]

These two events—our emergence as fully self-aware humans and our scaling up to literally take over the world—are consistent with the imperative for competition, but they don't answer the question: Why? All members of our human family (the genus *Homo*) had fashioned stone weaponry and utilized fire. Other members of the human family had also migrated out of Africa, yet only we arrived at the defining state of self-awareness.

The answer to this question flows from a central premise of this book, that there was a cognitive change, a shift in the interconnectivity of our minds, an adjustment in the way we perceived ourselves and the world around us. The two sides of our "internal conversation," our opposite minds, shifted ever so slightly to achieve a new balance. We became more fully two minds in one body and were better able to draw upon our collective ancestral intelligence, both male and female.

Prior to this shift, and for overriding purposes of survival, the left, male-influenced half of our minds had long maintained near-total dominance, but, in a greening period of plenty in Africa approximately 100,000 years ago, we have evidence of new behaviors that confirm innovation, artistic emergence, and a quantum move beyond the primitive status quo.

The new impulses were the recognizable creative/innovative features of the right, female-influenced half of our minds. This may seem counterintuitive, since the new creative impulses were instrumental in advancing the killing technology of projectile weapons: bow and arrow, as well as the possibility of velocity-enhancing atlatl hooks for throwing spears. However, this is only the first of the reconsiderations of gender that we will encounter as we explore this all-important central premise. Absent these breakthroughs in competitiveness, the cycle leading to our self-awareness may have been delayed indefinitely or emerged in a competing member of the genus *Homo*.

In this book we trace our 100,000-year journey of increasing scale and complexity in the binding of human minds to shared belief and purpose. Gender, at the scale of body and mind, informed every step of this evolution in unexpected ways. The eventual projection of our two "opposite" minds at nation scale formed our model of governance. Democracy, as well as the organization of corporate, military and cultural institutions in the third millennium, concludes the trajectory of our premise, which is organized in six chapters as follows:

1. **The Self-Aware Mind**

   We begin with our emergence from an animalistic state in the Middle Paleolithic Era, 200,000–40,000 years ago and our placement within the greater human family (the genus *Homo*). The dawning of our self-awareness is identified as the defining event for all life forms on Earth, and the violent competition for survival is recognized as the major "forcing'" factor in its emergence. The sources of our contemporary aversion to recognizing deadly violence as a central formative feature in our emergence are also addressed in this chapter.

   In the ultimate competitive event of the genus *Homo*, the most competitively evolved mind (*Homo sapiens*) confronted the most competitively evolved body (*Homo neanderthalensis*), and the outcome set the course of human history to this day. Although this confrontation occurred between the most similar members of the human family, there were differences of mind that proved to be a matter of life or death. Our self-aware ancestors prevailed and inherited the Earth.

2. Two Minds: One Body

As a foundation for understanding the evolution of our minds (our internal conversation and the wellspring of our differentness), we examine in detail the patterns of violence and early death as experienced by men and women in the Paleolithic Era. Forensic and archaeological evidence from this period and the subsequent violent transition from hunter-gatherer to agrarian cultures in the Neolithic Era provide validation for the formative role of violent competition in the development of our two minds.

Utilizing contemporary developments in the neurological sciences, we construct a profile of the two mindsets contending for "top of the genus." Profiles are based on known behaviors, skills, habitat, and strategy. In the end, *Homo sapiens* clearly displayed more right-hemisphere qualities and *Homo neanderthalensis* more left-hemisphere qualities.

Just as with the emergence of self-awareness, the determinant advantage for *Homo sapiens* was in the reduction of the male-influenced left-hemisphere dominance and greater integration of the female-influenced right hemisphere.

3. The Necessity of God

The inevitable product of the self-aware mind, in the face of an infinite universe, is the explanatory narrative. By explaining away the most frightening unknowns of death and an uncaring universe through the embrace of God, gods, or the supernatural, humankind went beyond the small bands held together by the biological bonds of blood and sex and scaled up to larger and more competitive belief-bound groups. We examine a couple of these sequentially larger and more competitive case studies, none more instructive than the civilization that arose in the Nile River delta 5,000 years ago.

The remarkable advancement of ancient Egypt toward the optimum organization of competitive and social life-forms, the superorganism, foreshadowed the trajectory of modern societies. The nature and utility of moral conduct, so meticulously interpreted in ancient Egypt, is explored, as well as why religion was the logical first, and still the most powerful, of the explanatory narratives. "Godless" states such as the former Soviet Union and, at a smaller scale, the New Atheists, eventually put forth their own explanatory narratives demonstrating that God is not a mandatory ingredient. However, over time it becomes clear that the relevance of any narrative is measured by the longevity and power of the bonding effect on the societies they inform.

The biological provenance of the explanatory narrative, a deep-seated, defensive imperative of the male-influenced hemisphere, is examined. The female-influenced hemisphere emerges as the primary interpreter of the explanatory narrative.

4. The Gendered Brain

Here is the heart of the argument. In Chapter Two, drawing on the work of the neurologist Louann Brizendine and the psychiatrist Iain McGilchrist, among others, as an overlay to the insights of the pioneering anthropologist Lawrence Keeley and others, we explore the basic implications of a new interpretation of the bihemispheric brain.

The central focus of Chapter Four expands this interconnection between the two sides of the brain, which is achieved primarily through a complex network of "hard" connections and "soft" influence. Our diversity and our individuality are established just before and after birth by the ways in which these interconnections and influences achieve a unique operational blend of maleness/femaleness in each of us via the universe of RNA. I have designated the unique balance between our minds as the *neurological gender ratio* (NGR). These are the two voices of "*Nature*" that engage the voice of "*Nurture*," our free will.

The importance of where we fall in the maleness/femaleness range of inclinations, our NGR, is illustrated by the life-or-death implications of that ratio when we face life's greatest threat: other humans. The ultimate test of our survival is the decision to trust or not to trust others. This determination is critical whether we are entrusting someone with guard duty or child care, or entering into a nuclear arms agreement. Femaleness tends to trust, maleness tends to distrust. Neither is good or bad, and either may be the correct choice in a given situation. Therefore, it is the essence of a contingent decision, a judgment call.

Humankind's viability as a species is enhanced by maintaining a range of inclinations in all men and women in the face of such dilemmas. Darwinian evolution is not going to deliver us to a "right" answer because there is no right answer. Sometimes yes, sometimes no—it depends. RNA, by creating neurological gender diversity has the effect of enhancing survival of the species in the face of such contingent decisions, as well as threats that will arise in an unknowable future. In short, the continuous variation, or spectrum, of maleness and femaleness is an inherent expression—including heterosexual, bisexual, and homosexual preference and transgender identity—of a resilient humankind.

Further observations in Chapter Four address the self-righting characteristics, as well as the grouping effect that is imparted through the DNA/RNA/"Two Minds" model. These dimensions of resilience in the face of future catastrophes and opportunities represent new concepts of *nonadaptive resilience*.

5. The Third Millennium Mind

Here we follow the evolutionary dynamic into modern times and find its expression in how we organize our societies, structure our corporate and military leadership, explore fields of knowledge, and compete through national and international sporting and athletic events.

The two great modern institutions emulating nature's "congress" of our minds (a continual open debate on the consequences of our changing environment and the corrective actions needed) are the models of democracy and science. Due to their underlying foundation in nature, they tend to exhibit an upwardly evolving and self-correcting behavior over time. Our Paleolithic legacy, expressed in reenactments in sport and media, and our clever assemblies of the total "DNA mind" in competitive corporate structures, are all folded back to examine their foundations in nature. The connectivity and concordance of our daily lives, interests, and concerns are more easily understood in this expanded evolutionary context.

The third millennium's more stable and secure world, with its reconciliation of the conflicts of the major world powers down to the scale of skirmishes and tugs of war over small islands and peninsulas, has facilitated the flourishing of the creative, right-brain segment of the population and, in the West, an atmosphere of increasing tolerance across the spectrum of human uniqueness and variance from "normal." This cultural/political shift and its far-reaching implications are examined in Chapter 5.

6. **The Revelation**

The final chapter reveals a stunning finding from 2014, an unanticipated forensic insight in support of the central premise of this book on the threshold of its publication, as well as the discovery, in 2015, of a crossover validation in epigenetic biology. The author's conclusions and observations are encapsulated in ten interlocking concepts set forth for the first time here. The implications of these observations range across three scales:

- AT THE BIOLOGICAL SCALE of our mindset there are two evolutionary forces, each operating in opposite directions: DNA/natural selection converging on the most resilient physical attributes for our species, and RNA/neurological diversity imparting a range of worldviews such that each individual has a unique perception of the world and the intentions of others. DNA increases the resilience of human individuals in existing environments, while RNA increases the resilience of the species in unknowable future environments.

- AT HUMAN SCALE, the emergence of self-awareness emanated from the creative/opportunity side of our natures, the right-brain (femaleness) as opposed to the left-brain (maleness) characteristic of protecting and defending the status quo. Our left brain, for reasons of survival, was dominant in prehuman times. The transition to the increased influence of the right brain marked the emergence of humankind and has continued to increase in influence over time. The two hemispheres of our brain are the indispensable and interdependent halves of who we are—our natures.

- THE CONCLUSION AT THE LARGEST SCALE is that our 100,000-year journey has been a sequence of three global evolutionary revolutions to higher levels of density and complexity. All three have a biological provenance and are transitions toward the highest form of competitive and social order: the superorganism.

Despite our strong desire to categorize and normalize all the variables of life, the most inaccurate term that can be applied to an individual human is to say that he or she is "normal." There is, even between identical twins possessing identical DNA, a deep level of uniqueness in the way each perceives the world and, based on that perception, acts. That uniqueness, a universal human characteristic, is intertwined with gender in ways that require a reconceptualization of its role in all our lives.

Shakespeare, in Polonius's advice to Hamlet "To thine own self be true," recognized this underlying reality, and it turns out that we are each unique in the absolute sense of the word. The purpose of that uniqueness is no less than to assure that humankind will survive in times of catastrophe and thrive in times of opportunity. The common bond of humanity is that we are all born with different ways of being in this world, and—in the context of the modern world—the nation that can free the individual potentials of its citizens while achieving international and domestic security will have the greatest chance to survive and thrive.

CHAPTER ONE

The Self-Aware Mind

OUR EVOLUTION MAY BEST BE UNDERSTOOD AS AN INHERITANCE from the previous 125,000 generations of the human family, or, more accurately, the genus *Homo*. Their victories and failures over the past two and a half million years informed the final "pathway" leading to our bihemispheric brain, which, although similar in all mammals, is uniquely realized in us.

This brain comes to us preloaded with survival/reproductive knowledge that is both a background (unconscious) and deliberative (conscious) intelligence. We live in the world of our minds—our internal environment—and also in the world around us—our external environment. These two worlds evolved in parallel and relative isolation over the billions of years of the development of life on Earth until they were merged in our species.

Something happened. Our internal world, the world of our minds, crossed a line: we became aware of our ability to mold the external world. With this awareness, our internal and external worlds joined in a process of coevolution in which human minds and actions, more than nature, dramatically change the external world. Meanwhile, the internal world of our minds struggles to compete in the shifting realities of a world that we are constantly changing.

THE DAWNING OF SELF-AWARENESS

The merging of these two worlds, our self-awareness, is a near-miraculous event that occurred approximately 100,000 years ago in Africa following a period of extreme drought and volcanic winter.[1] Reduced to near-extinction, with our genetic profile pointing to a dangerously low population of approximately

10,000 birthing couples, we entered a period of climate stability and expanding biodiversity. We now have evidence from the Sibudu and Blombos caves of South Africa that during this flowering of the African garden, our ancestors advanced the technology of the Middle Paleolithic in ways not achieved by any other member of the genus *Homo*.[2] Bone arrowheads confirm this strategic advantage in combat, compound adhesives for hafting spear heads to shafts testify to an advancing knowledge of material properties, and the retrieval of bone needles from the site evidences an ability to tailor clothing and thereby adjust to climate extremes as required. But the most important innovation was that of projectile weaponry, a breakthrough that was further substantiated in findings at Porc Epic Cave,[3] some 3,000 miles north in modern-day Ethiopia and the launching point for further ancestral migrations north.

The essential advantage of removing the combatant from direct physical contact, and thereby potential harm, through the airborne projection of lethal force is a marker in the advancement of weaponry. From the hand-held thrusting spears of *Homo neanderthalensis* to the drone pilot sitting in Tampa, Florida, "flying" over Afghanistan and firing Hellfire missiles, we witness the arc of evolution in the projection of force. Therefore, it is the projectile points of the spear and arrow and the projectile assemblies of the spear throwers (atlatls hooks) and bows that are most relevant—the most determinant.

Projectile weaponry, by creating a durable advantage in the search for food and destruction of enemies, leads directly to abundant resources and reproductive acceleration. Generation by generation, the expanding interface with the external world continually rewards this pattern of innovation and thereby increases the number of these mindsets.

The resulting complex advancements of *Homo sapiens* stand in stark contrast to their (and therefore, our) closest relative, *Homo neanderthalensis*, who, over a 300,000-year presence and a 200,000-year dominance in Eurasia, remained in a state of relative evolutionary stasis. That is to say, they did not significantly advance their organization in small, isolated bands or their basic sharpened-stone and thrusting-spear technology. The recent evidence of their more varied diet, pierced sea shells for adornment, use of red pigment, and adapted leather-shaping tools[4] does not rise to the level of the expanded scale, organization, and socialization of human groups and advanced cultural expressions that characterized *Homo sapiens*.

Based on everything we now know, a strong circumstantial case can be made that *Homo sapiens* entered a natural selection feedback loop of innovation/ competitive edge/accelerated reproduction which created, through natural selection, the ultimate advantage over all other life: the self-aware mind. This was essentially a shift from thinking in limited terms of survival to thinking in terms of context and possibility.[5] From this foundation our ancestors initiated the ultimate comeback from near-extinction, launching an outward migratory expansion into the Near East, Central Asia, Europe, and beyond. Our species became the dominant life form on every habitable landmass on the planet and emerged as the sole survivor of the genus *Homo*.[6]

THE HUMAN FAMILY

You might ask what distinguishes the members of the genus *Homo* from our previous ancestors and how one qualified to get into this club. Two of the basic qualifications were the ability to control fire and the ability to fabricate and use stone tools. Among the items in the typical Stone Age tool kit were sharp-edged, hand-shaped stones for cutting and scraping, which were also used as spear points and knives. Therefore, the family of genus *Homo* constituted the only armed inhabitants of this critical transition period, the Middle Paleolithic Era (200,000–40,000 years ago).

For the purposes of our discussion we include the six members of the genus listed in the Smithsonian's *Human Species Diagram*, which was developed as part of the Human Origins Project in 2010 and published in *Smithsonian* magazine. In addition, although this is a recent finding, there is the upgraded status of a currently designated *'hominine'* discovered in Southern Siberia's Denisova Cave: *Homo denisovan*.[7] This addition to the genus *Homo* (a working assumption) flows from both the strong initial scientific consensus on the authenticity of *Homo denisovan's* richly detailed genetic profile and predictive value in the reconsideration of the nature of human emergence. Table 1.1 constitutes a listing of the human family tree at the start of *Homo sapiens* migration/dispersions out of Africa and across the planet.

The accurate grouping of the genus *Homo* is a key to understanding our human family, because of the many misleading names that have been used in the popular press and/or locales of discovery. For instance, we are often referred to as Anatomically Modern Humans (AMHs) while in the Paleolithic time periods "Cro-Magnon" and "Aurignacian" are often used rather than *Homo sapiens*. *Homo erectus* is often referred to as the "Java Man." What is important to know is that these ancestors shared a set of skills at the threshold of self-awareness. They were all, to lesser and greater extents, "knocking on the door," which confirms a commonality in their rise toward a higher consciousness.

Another characteristic central to this family was a drive to explore, to acquire land, and, while in possession, to be fiercely territorial. To this day, our urge to acquire territory, to harness fire, and to create weapons remains at the center of our culture.

The assumption that the members of the genus lived in isolation and had little or no interaction has recently been proven incorrect. The DNA analysis of *Homo denisovan* and reconstruction of the DNA of *Homo neanderthalensis* has opened a window on these relationships. Both DNA breakthroughs were informed by the research of Svante Pääbo at the Max Planck Institute of Evolutionary Anthropology in Leipzig, Germany.[8] The findings have been shocking: the genetic substantiation of interbreeding between *Homo sapiens* and at least two other members of the Genus, *Homo neanderthalensis* and *Homo denisovan*. A brief overview of the process by which we have come to know our evolutionary history places these findings in perspective.

ORIGINS OF THE HUMAN FAMILY

Ours was but one of the multiple waves of the human family who, in response to crisis or opportunity in their drive for survival, migrated out of Africa. These migrations, more accurately called dispersions, were separated by tens or even hundreds of thousands of years, resulting in multiple branches of a common ancestor evolving in different natural settings. The differentiation of these groups became so significant that when one splinter group (say, *Homo sapiens*) came into contact with another (say, *Homo neanderthalensis*) approximately 458,000 years after they split off from a common ancestor, there was a chasm of difference— physical appearance, language, settlement patterns, and culture. A significant difference in their perception of the world also separated *Homo neanderthalensis* from *Homo sapiens*.

Over 170 years ago, Charles Darwin first noted the process of natural selection and descent from a common ancestor in the twenty-five species of small birds on the Galápagos Islands whose lineage was eventually traced to a single bird species on the mainland of Chile. Referred to as Darwin's finches, their unique beaks were different from island to island and were specifically shaped to the feeding opportunities in their setting. These observations were instrumental in changing Darwin's mind on the conventional notion of the stability of species and reinforcing the idea of transmutation of species. This was to be a central insight in his revolutionary Theory of Natural Selection, first published in *Origin of Species* in 1859. More to the point of our interests, Darwin, in his 1871 blockbuster *Descent of Man*, asserted that the descent of all humans was from a single ancestor (monogenesis) and that that ancestor would have been from Africa.

In 1987, one hundred and sixteen years after Darwin's prediction, Rebecca L. Cann, Allan C. Wilson, and their colleagues at the University of California, Berkeley put forth a bold and all-embracing theory on the origin and rise of *Homo sapiens*, which was framed in biblical simplicity and supported by a scientifically rigorous genetic validation.[9] In their ground-breaking paper, they convincingly

Table 1.1. The genus *Homo* at the time of
the *Homo sapiens* Migrations Out of Africa: 60,000 YBP

Homo habilis	**extinct:** possible ancestor of *Homo erectus*
Homo erectus	**extinct:** possible ancestor of *Homo heidelbergensis*
Homo heidelbergensis	**extinct:** probable ancestor of *Homo neanderthalensis + Homo sapiens + Homo denisovan*
Homo floresiensis	**alive** in Indonesia/East Asia
Homo denisovan	**alive** in Russia
Homo neanderthalensis	**alive** in Europe
Homo sapiens	**alive** *(dispersing north into Central Asia)*

Note: The time period of modern human emergence (200,000–40,000 YBP) is referred to as the Middle Paleolithic or Middle Stone Age by anthropologists, who mark this period by the use of stone tools. Geologists refer to this period as the Middle or Upper Pleistocene and mark it by the coming and going of the Ice Ages and related glaciations.

argued that all of humankind descended from a single species of modern human, *Homo sapiens*, and that their common heritage, our lineage, was traceable back 200,000 years to Africa—monogenesis confirmed!

Their argument was built on the fact that the human mitochondrial DNA (mtDNA) is a powerful generational marker, passed from all mothers to daughters. Utilizing this insight, they traced the entire maternal genealogy of our species back to a common ancestor for that gene (a phylogenetic tree of humankind). In 2001, the more complex male counterpart to mtDNA (the Y chromosome, or Y-DNA) was finally analyzed, confirming the African ancestry and approximate time period.

The most controversial aspect of their theory related to the nature of the outward expansion from Africa, which they dated to 100,000–50,000 years ago and which was captured in their statement that modern humans (*Homo sapiens*) "replaced" all other archaic humans, *Homo neanderthalensis* being the only member of the genus known to be alive in that time frame.

This raised a wide range of challenges from the scientific community, with many insisting that interbreeding must have taken place and/or that other specific archaic humans (*Homo erectus* being one candidate) could have been around from an earlier migration to Eastern Asia and have been assimilated through interbreeding. However, all of these alternate scenarios did not explain the low genetic diversity that characterizes humankind today and which was completely consistent with the theory put forth by Cann & Wilson, et al.

THE NATURE OF NATURAL SELECTION

The key controversy that continues to this day is the nature of competition that occurred between *Homo sapiens* and *Homo neanderthalensis* and possibly other members of the human family in this process of replacement. Was it a process of nonviolent competition for resources; sporadic conflict, war, genocide, and cannibalism; or did *Homo sapiens* just enter empty caves after *Homo neanderthalensis* or other members of the genus *Homo* had departed?

The years 2009–2010 were a pivotal time for consideration of these two very different members of the genus *Homo* and the nature of their Middle Paleolithic family reunion. Two books specifically addressing the *Homo sapiens/Homo neanderthalensis* dynamic were published: *The Humans Who Went Extinct: Why Neanderthals Died Out and We Survived* by Clive Finlayson (2009)[10] and *CRO-MAGNON: How the Ice Age Gave Birth to the First Modern Humans* by Brian Fagan (2010).[11]

In addition to these two books, both aimed at a broad readership in popular science, the ambitious David H. Koch Hall of Human Origins opened at the Smithsonian National Museum of Natural History in Washington, DC, in March 2010, with National Geographic publishing the companion book *What Does It Mean to Be Human?*, by Richard Potts the director of the Smithsonian's Human Origins Project, and Christopher Sloan, an expert in paleoanthropology at National Geographic.[12]

All three books and the Smithsonian exhibit dismissed violent confrontation as characteristic of the *Homo sapiens* dispersion, with several authors attributing a major, even a stand-alone, role for climate change in the fate of *Homo neanderthalensis*. As if to add an exclamation point to this avoidance of competitive violence as a fundamental evolutionary force, *Scientific American*, in September 2014 published its Special Evolution Issue: How We Became Human, titled *Evolution: the Human Saga*. Climate change, in the form of the "rapid climate cycles" that characterized the Africa of our ancestors, is credited in several of the articles in this issue as the driver of evolution. A key argument is that accelerated evolutionary change occurs at times of great climate upheaval (a well-founded assertion, not in dispute).

However, the conclusion drawn was that in times of scarcity of food and water our forebears' key to survival was their "varied diet" as opposed to the inflexibility of other prehumans who were vegetarian and "eventually became extinct."

There is another more brutal and realistic explanation. In times of major climate disruption, drought and the loss of one or more growing seasons exponentially increases the competition for food and water. Die-offs in the most extreme events and their competitive aftermath can range from 50 percent to over 90 percent of species. A small segment of the strongest and most aggressive will often prevail. When the good times return, this fraction of the original population will survive as the dominant version of the species or, under certain conditions and in certain time frames, constitute a new species.

I agree that there may be a "dietary factor" at play, since humans kill and consume other life forms for their survival, even going to the extreme of cannibalism when starving, as opposed to pastoral vegetarians. This, I would argue, is the factor linking diet to human survivability in such hypercompetitive environments.

The editor of the *Scientific American* Special Evolution Issue specifically states that "fresh insights" call for the revision of "virtually every chapter of the human saga from the dawn of humankind to the triumph of *Homo sapiens* over Neanderthals and other archaic (ancestral) species." Completely to the contrary, the evolutionary imperative of competition (as violent as necessary to prevail) is the logical driving force in the emergence of *Homo sapiens*. The occurrence of accelerated evolutionary change at times of climate upheaval—the centerpiece of the *Scientific American* issue—is, I would argue, totally consistent with this competitive reality.

Nonviolent scenarios are completely contrary to the earlier assertions of scientists on the other side of the argument, first among them Jared Diamond, author of *The Third Chimpanzee* (1991), *Guns, Germs, and Steel* (1997), and *Collapse* (2005).[13] More recently, Richard Pinker, in *The Better Angels of Our Nature* (2011),[14] Edward O. Wilson, in *The Social Conquest of Earth* (2012),[15] and Christopher Stringer, in *Lone Survivors* (2012),[16] have all weighed in on the side of a violent provenance for early humankind.

This is more than a difference of opinion. The 2009–2010 four-way characterization of nonviolence and the more recent 2014 *Scientific American* characterizations are part of a larger pattern that deserves elaboration. What could be the source of these nonviolent interpretations from a large segment of the natural sciences community in the face of accumulating biological evidence and arguments to the contrary?

In fairness, there are deeply traumatic events in the recent history of our human community that have contributed to the persistent denial of violence and racism as biological features of our mindsets. First among these events: Adolf Hitler's Germany and his genocidal pursuit of a Master Race. Many of us still want to believe that Hitler was an exception and that no such dark potentials exist in the mind of humankind. However, an uncomfortably similar American and British point of view was espoused in the same time frame.

Eugenics (the hereditary improvement of humans by selective breeding) was a popular social movement in America and Europe in the early twentieth century and was the subject of international conferences in London in 1912 and in New York in 1921 and 1932.[17] Prominent members of the movement included Leonard Darwin (the son of Charles Darwin), Winston Churchill, and Alexander Graham Bell.

The most severe U.S. public policy enforcing this view was the mandated sterilization of those found to be "unfit." Nazi war criminals later cited this movement at the Nuremberg Trials as a source of their policies and excuse for their actions. Not only eugenics but any version of a biological basis for human behavior was swept up in the unanimous condemnation of Nazi crimes against humanity.

The United Nations adopted, in 1948 and 1978, declarations affirming the *equality of all humans*, which have undoubtedly helped humanity and the cause of freedom over the ensuing years. However, the UN also adopted the 1986 Seville Statement on Violence, which enlisted a group of leading scientists to affirm that violence by humankind had no biological basis.

The declaration ended up as five statements:

1. It is Scientifically Incorrect to say that we have inherited a tendency to make war from our animal ancestors.
2. It is Scientifically Incorrect is to say that war or any other violent behavior is genetically programmed into our human nature.
3. It is Scientifically Incorrect to say that in the course of human evolution there has been a selection for aggressive behavior more than for other kinds of behavior.
4. It is Scientifically Incorrect to say that humans have a "violent brain."
5. It is Scientifically Incorrect to say that war is caused by "instinct" or any single motivation.

Stephen Pinker has correctly characterized these statements as an example of a *moralistic fallacy*, that is, projecting our moral values on biological phenomena. A wide range of findings has subsequently converged to support an underlying biological/behavioral dynamic in humans. Now, over thirty years later, the Seville Statement is entering a long overdue period of reconsideration and revision.

The furious objections to E. O. Wilson's book *Sociobiology*,[18] published in 1975, confirmed that a similar collective nerve had been struck in his characterization of the human mind and masterful articulation of the argument that human social behavior is biologically based—a product of evolution.

Certainly his sweeping and provocative statements, such as: "Free will is an illusion," did little to soften the message! A brilliant "partner in crime," Wilson's fellow Harvard professor B. F. Skinner, penned a few provocative lines on the subject, such as:

> An organism behaves as it does because of its current structure...a person's genetic and environmental histories. What are introspectively observed are certain collateral products of those histories.[19]

These early statements came across as absolutist, and led to charges of biological determinism, in the sense that they suggest that humans are directed in their behavior, without participation in the decisions that shape their lives. While independent validations for many of the arguments supporting such a biological basis have been widely accepted, the passionate objections and legacy of the opposition remain firmly intact. The contemporary overstatements of the role of climate change in the Middle Paleolithic Era are, in my opinion, continuing expressions of this aversion to recognizing the full significance and formative role of competitive aggression and violence.

My argument is in support of a biological basis for our mindset, which is neither violent nor nonviolent but possesses two points of view, including both aggression and empathy. It is important to note that these two states of mind are deliberative in nature and, across the species as a whole, operate in a range from subtle inclinations to "stubbornly intransigent" will. The key distinction between us and all other life is self-awareness, whereby we can observe from a distance the consequences of these inborn predilections and make adjustments. In fact, if there weren't such a version of human free will, the upwardly evolving force of natural selection (upward, in this case meaning increasing complexity and consciousness via random processes) would be greatly diminished. In other words, we are each able to exert a greater or lesser degree of "free will." We, in real time, are active participants in the internal deliberations of our minds, which have been informed by the survival pathways of our ancestors.

Setting aside the pacifist versions of the demise of *Homo neanderthalensis*, we will attempt to build a comprehensive framework of human behavior in the Middle Paleolithic Era. We begin with the departure of *Homo sapiens* from Africa and conclude with the forensic and behavioral evidence framing the nature of these two competitors in the last days of *Homo neanderthalensis,* the Paleolithic Commander of Eurasia.

Figure 1.1 is the map of *Homo sapiens* pathways out of Africa and dispersion to all habitable landmasses. In addition, the dating convention of this book is established. Rather than "years ago" and the various conventions of BC, AD, and BCE, et cetera, all dates will be expressed as years before present: YBP. Thus, 200–130k YBP means 200,000–130,000 years before present, HS = *Homo sapiens*, and HN = *Homo neanderthalensis*.

Note: there are a number of significant dates that are the subject of ongoing debate within the scientific community, but those differences have no impact on the broader concepts being covered herein. For simplicity and narrative flow, the following dating convention has been adopted for the telling of our history:

1. Common ancestor of HN and HS = 500,000 YBP
2. HN emerges as a species in Europe = 300,000 YBP
3. HN dominates Europe/Central Asia = 200,000 YBP to 40,000 YBP
4. HS emerges as a species in Africa = 200,000 YBP
5. HS transitions to self-awareness = 100,000 YBP
6. HS migrates out of Africa = 60,000 YBP
7. HN/HS conflict = 42,000-40,000 YBP
8. HN extinct 40,000 YBP

The intense interest of the media and the general public in the fate of *Homo neanderthalensis* and their disappearance from the fossil record 40,000 YBP is understandable because they are by far our closest relative to have walked the earth; 99.7 percent of our DNA genetic heritage is identical, and we share a common ancestor dating back 500,000 YBP, or approximately 25,000 generations.

As a comparison, our closest living relative is the chimpanzee, which, in addition to its human-like and adorable antics in childhood, matures into the only member of the animal kingdom to form all-male raiding parties to attack and kill members of their own species who they perceive to be encroaching on their turf. Chimpanzees possess a hyper territoriality that is clearly shared by their cousins: us. However, even with these similarities of behavior, our common ancestor is much further back (at least 6 million YBP, or approximately 300,000 generations), and our DNA genetic heritage is only about 98.8 percent similar.

[Author's note: this high degree of equality between our DNA and that of the chimpanzee only serves to underline the staggering power of RNA to differentiate our respective species].

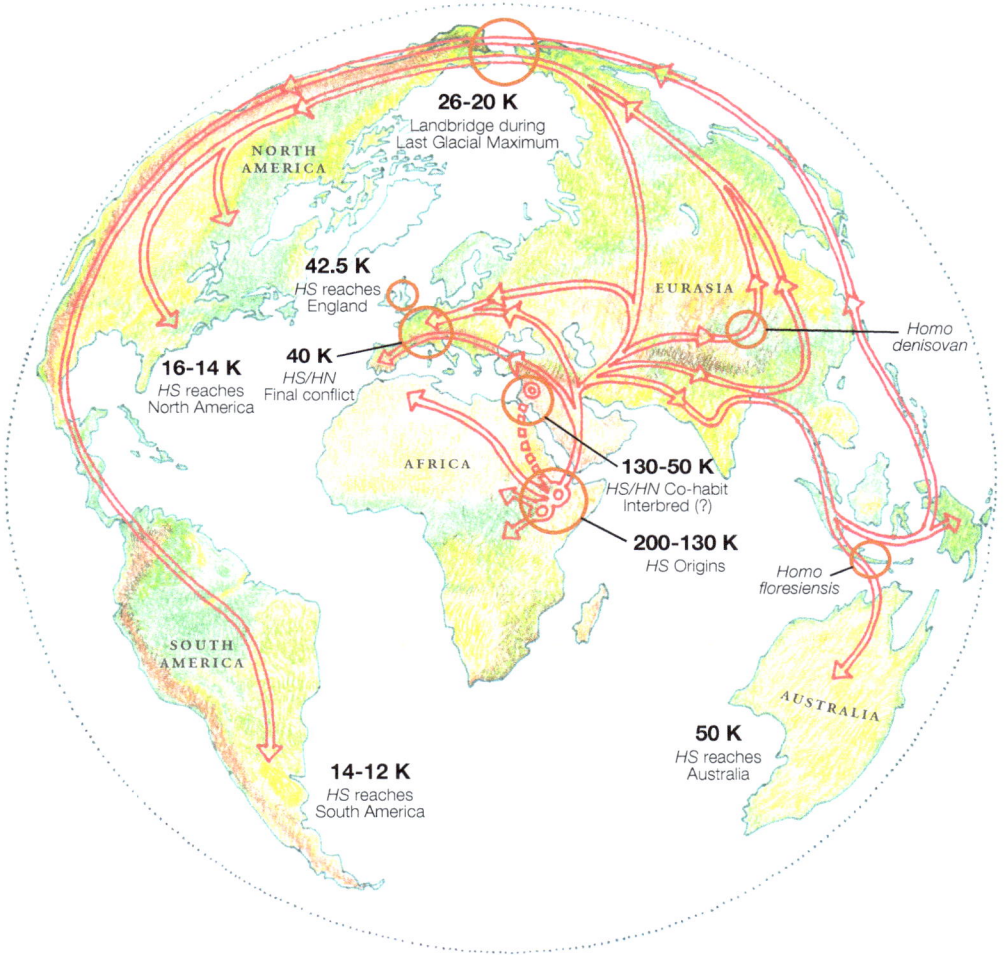

Figure 1.1. *Homo sapiens* Origin and
General Dispersion to all Habitable Land Masses

HS= *Homo sapiens*, **HN** = *Homo neanderthalensis*
40 K = Forty thousand years before present

(Note: Geographical Relationships Modified by Author for Clarity of Dispersion Pathways)

A generally accepted theory is that we share a common ancestor with *Homo neanderthalensis* and that our separation and genetic divergence result from an early and partial out-migration from Africa of that ancestor, creating a "Eurasian branch" that evolved into *Homo neanderthalensis* ±300,000 YBP, with the remaining "African branch" evolving into *Homo sapiens* ±200,000 YBP. When we dispersed out of Africa 60,000 YBP, it was the first step on the path to a reunion with our European cousins.

Approximately 25,000–15,000 years before the large-scale contact in Eurasia, there is evidence to support a relatively local occurrence of cohabitation by *Homo sapiens* and *Homo neanderthalensis* at Manot Cave in what is now Israel.[20] There are indications of interbreeding that would have occurred in this apparently failed outward migration between these more closely related "cousins" who were still in the same neighborhood and had not evolved to the level of cultural and genetic difference that would be encountered at the comparatively dark and frozen northern latitudes of Eurasia tens of thousands of years later.

My argument begins with an agreement, an agreement with Jared Diamond and others who originally recognized the Eurasian incursions as the defining ancestral confrontation, foreshadowing the extinction of all the remaining members of the genus *Homo* and protohumans on Earth. This was the threshold moment, the key large-scale trial of the two top brain-body combinations of the genus *Homo* in a competition that was eventually won by the innovative, strategic and aggressive mix that was uniquely *Homo sapiens*. I will argue that the defining competitive edge was one of cognition, an essentially *Homo sapiens* way of seeing the world, which made a life-or-death difference in that remote time and place, and that this shift in cognition continues to shape us to this day.

The separation of the two protagonists by 458,000 years from a common ancestor and the different environments of Eurasia and Africa led to an array of differences. Table 1.2 provides an intial comparison.

HOMO NEANDERTHALENSIS

A major advantage enjoyed by *Homo neanderthalensis* was the total domination of their range, having eliminated all human or prehuman competitors across Europe and into Central Asia. Although they appear to have arrived 300,000 YBP, their period of dominance was approximately 200,000 years. They were able to settle into an intimate alignment with flora, fauna, climate, and migratory patterns that provided their subsistence (primarily bison, red deer, horse, etc.). They were consummate hunter/stalkers, since their fire-hardened thrusting spears required close proximity to an animal for a kill.

Table 1.2. Top of Genus and Macro-Scale Comparisons

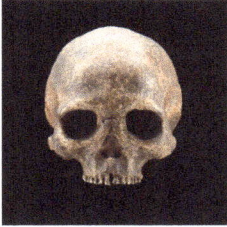

Homo sapiens		Homo neanderthalensis
(also referred to as "Cro-Magnon")		*(also referred to as Neanderthals)*
tall, thin, gracile	**stature**	short, barrel-chested
high flat forehead	**facial**	low, sloping forehead
rounded shape, no heavy brow	**skull**	square shape, heavy brow
lighter	**weight**	heavier
1,345 cu. cm.: average	**brain size**	1,450 cu. cm. (larger): average
projectile (from a distance)	**weapons**	thrust weapons (close range)
endurance, maneuver	**physical advantage**	brute strength in close
hilltops, defensible, with good visibility (strategic)	**settlements**	near water and food source (convenient)

When we consider the relatively brief stand of the North American Indian and the Inca, Aztec, and Maya in South America, which ranged around 18,000–8,000 YBP, we can only begin to grasp the significance of the 200,000-year reign of *Homo neanderthalensis* in Eurasia, a period thirty-three times as long as recorded history!

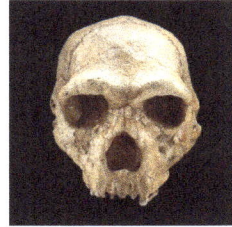

Because *Homo neanderthalensis* was primarily a meat-eater and killed large mammals in a singularly risky and time-consuming method, it has been noted (given the carrying capacity of the landscape) that the entire population was very thin on the ground, probably never exceeding 10,000–15,000. They remained in a state of relative cultural stasis for hundreds of thousands of years with very little progress or innovation, especially when compared to *Homo sapiens*. Although recent findings point to a more varied diet and possibly a larger population, their isolated bands, low socialization, and low competitive pressure within their range tend to support such a lack of cultural advancement.

Figure 1.2 provides comparisons of the two anatomical cores, with strategic advantages and disadvantages of each. Immediately upon examination of the heavy skeletal armor of *Homo neanderthalensis* rib cages, we recognize one of the most powerful musculatures ever developed in the human family, exemplifying a massive investment of evolutionary capital in brute strength. The history of their physical confrontations with man and beast are recorded in the high frequency of their broken and healed bones, a condition that is not seen in any other member of the genus *Homo*. In short, they had evolved to succeed in close-contact, brute-strength competition at cold northern latitudes, and no other member of the human family could defeat them one-on-one on these terms.

A striking facial feature of *Homo neanderthalensis* would have been the very large size of their eyes compared to ours. This, combined with a distinctive bulge at the rear of their skulls, called the "bun" (location of the visual processing center of the brain), suggests a vastly superior evolution in eyesight. This may have been expressed as higher resolution or better distance vision, but most likely would have been expressed as superior night vision, much like night vision goggles worn in modern-day combat. This advantage would have been of great value for hunting and stalking in the shorter days at these darker latitudes, so unlike their North African homeland.

In addition, as the first of the genus *Homo* to survive and persist at these northern latitudes, they became supremely adapted to the cold. Their body shape (short and stocky) created a thick center-body core that was almost entirely enveloped in large, expanded rib-cage "armor." Although these physical traits restricted movement, they protected vital organs during face-to-face conflict. These traits also reduced body surface area and therefore provided *Homo neanderthalensis* with much greater insulation than their earlier tall, slender African body type that gives up body heat more quickly.

Contrary to the impression that seems to be emerging here, there were some familiar genetic and anatomical features of *Homo neanderthalensis*:

M1CR gene: specifically related to pale skin and red hair.

Hypoglossal Canal: the nerve pathway connecting the brain and tongue, necessary for speech.

Thoracic Vertebral Canal: necessary for coordinating breathing lungs and diaphragm for articulated speech.

FOXP$_2$ gene: for language and expressive control of facial muscles.

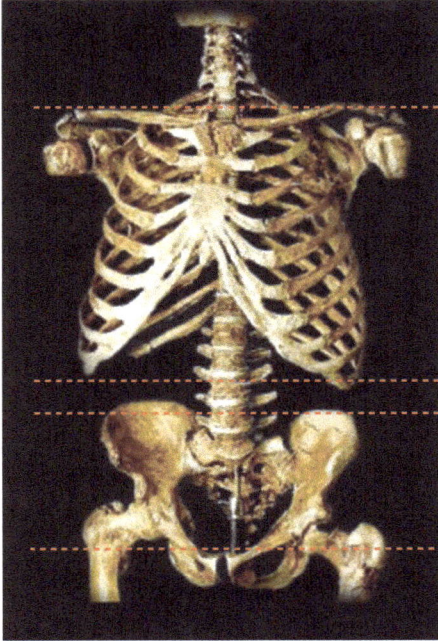

Homo Neanderthalensis

Massive upper body rib cage
Protection of vital organs

Low center of gravity/ High bone mass

Anatomy, musclulature, and proportion
support retention of body heat

Small exposure of mid-section

Large pelvic bone enclosure
with legs at wide stance for
bracing against impact

*(Note: Tradeoffs to gain 'armor' and
brute strength result in reduced range of
movement and long distance endurance).*

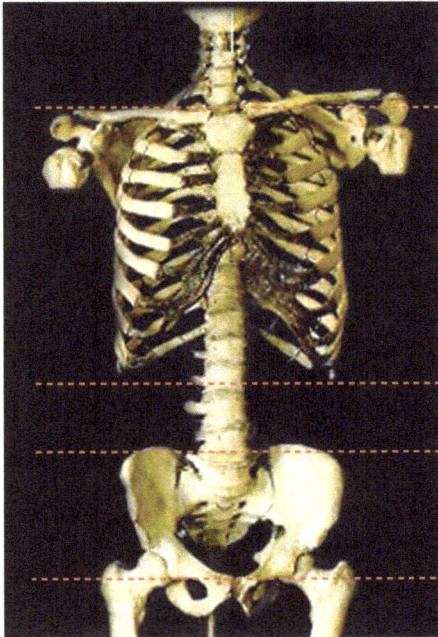

Homo Sapiens

Light-weight upper body rib cage
Vertical/ Upright Proportion

High Center of Gravity/ Lower Bone Mass

Anatomy, musclulature, and proportion
support endurance, long distance running
and rejection of heat from body

Wide exposure of mid-section allows
bending and rotational flexibility

Smaller and lighter pelvic bone enclosure
with legs at relatively narrow stance;
long legs facilitate maneuver and running

*(Note: Tradeoffs to gain maneuver, speed, and
endurance result in less brute strength and
greater exposure to injury. Clear advantage
in throwing projectiles is gained).*

Figure 1.2. Anatomical Core Comparisons with Strategic Implications Noted

Far from the inarticulate grunts of "Neanderthals" in the familiar cartoon version, *Homo neanderthalensis* was capable of articulated speech and was more recognizable as a human than typically portrayed. Although we may find it an appealing notion, that these two most intellectually advanced and physically superior members of the genus *Homo* had so much in common, having the same genetic heritage, the same need for territory, food, water, and shelter was a formula for deadly conflict when close relatives must survive in the same territory. Known as the *competitive exclusion principal*, or Gause's law, this phenomenon is well described in the early work of W. L. Brown and E. O. Wilson.

HOMO SAPIENS

Homo sapiens left Africa 60,000 YBP in possession of superior weaponry, language, and a fundamentally different cognitive capability that fueled their successes as they moved north and into Central Asia. They possessed the competitive edge of self-awareness and brought with them hunting and military strategies beyond the capabilities of their opponents.

A key evolved physical difference in *Homo sapiens* was the higher rib cage, which, although exposing more vital organs of the mid-section, created a greater length of unrestrained spine and allowed much greater rotational and bending flexibility for maneuver and, most important, for the throwing of projectiles. The two opponents—*Homo sapiens* versus *Homo neanderthalensis*—represented the classic boxing matchup: the agile, fast-reaction "boxer" throwing punches from a distance versus the close-in, brute-strength "brawler."

As opposed to the small scale of the *Homo neanderthalensis* band and their focus on large mammals, the incoming *Homo sapiens*, in larger collaborative and more socially integrated groups (which increased efficiency and reduced risk) pursued a broad range of small animals, fish, and birds, as well as big game. This approach supported a higher density of population for a given area of land, and the general consensus is that they arrived with or built up to superior numbers over time.[21]

While *Homo neanderthalensis* had clearly been forged of the toughest steel, their African cousins, although from a common ancestor, had followed a different path. Neither cold-adapted by body type nor physically stronger, *Homo sapiens* arrived on the field of competition with evolutionary adaptations and cultural innovations that defeated those physical advantages. Their evolutionary capital included a strategic investment in the nonmuscular realm of the bihemispheric brain, and from this seemingly subtle difference, *Homo sapiens* had an innovative and defining advantage.

THE INVISIBLE ADVANTAGE

At this point, we have considered the physical baseline and competitive characteristics of our *Homo sapiens* ancestors at the threshold of Eurasia and their final testing against *Homo neanderthalensis*, the strongest opponent they faced within the genus *Homo*.

As a good journalist, we have developed the what, where, and when of the story but have yet to address the how and why at the level of their very different mindsets.

Our assertion of a cause-and-effect relationship between a killing technology, and accelerated reproduction, on the one hand, and the emergence of our self-awareness on the other, does not explain how such a transformational change became a permanent feature of our reality. To gain such an understanding we need to consider the biological pathways of perception and consciousness at this last major threshold of our animalistic past.

Homo sapiens and *Homo neanderthalensis* possessed near identical bihemispheric brains, two halves of a whole, massively interconnected to allow the transfer of information and influence between the hemispheres. And yet they perceived two very different worlds, a difference that turned out to be a matter of life or death. Any understanding of the emergence of *Homo sapiens* and our eventual survival beyond all other members of the genus *Homo* requires an understanding of why the bihemispheric brain is divided and how it has come to be uniquely expressed in us.

To this end, in Chapter Two we analyze the forensic evidence, artifacts, strategies and behaviors of these contenders in order to "reverse engineer", to infer from these givens, the nature of their minds at work. In addition, we take advantage of current advances in the mapping of the two hemispheres of our brains (two minds: one body), to gain a perspective on their, and our, different perceptions of the world at that time and place: 42,000-40,000 YBP in Eurasia and, of particular interest, the lands now comprising Western Europe.

Two Minds, One Body

WE HAVE AN UNEXPECTED SOURCE OF INFORMATION ON THE NATURE of competition and cognitive advantage as they played out in the Middle Paleolithic. Since *Homo sapiens* took exclusive Eurasian ownership only 40,000 YBP—a short time ago on the evolutionary timescale—the brain of *Homo sapiens* that confronted *Homo neanderthalensis* is essentially the same brain that is reading the words on this page.

For this reason, we are able to draw on established patterns of human behavior in such confrontations as well as on the recently developed mapping of the characteristics of our bihemispheric brains. These advances in the behavioral and neurological sciences provide a new perspective on the Middle Paleolithic and the challenges our ancestors faced in their competition within the genus *Homo*. In addition, the later, violent replacement of these hunter-gatherers by Neolithic agrarian settlers in Europe 8,000–6,000 YBP provides a striking parallel: a repeat performance of an incursion by an advantaged aggressor moving across the same lands.

We draw here on three highly relevant explorations of human intention and cognition. Lawrence H. Keeley, in his well-documented *War Before Civilization: The Myth of the Peaceful Savage* (1996),[1] provides a compelling analysis of the inherent presence of human-on-human violence in hunter-gatherer societies, while Jared Diamond, in *Guns, Germs, and Steel: The Fates of Human Societies* (1997)[2] expands on the nature of our long history of incursions as advantaged aggressors into disadvantaged indigenous populations. The unique circumstances of early death that emerge from this analysis underline the formative role of violent competition in the creation of two very different survival pathways for men and women. Natural selection served to concentrate these two viable outlooks, which are a legacy of ancestral guidance embodied in our minds.

Finally, we turn inward to consider the cognitive difference that existed between our two Middle Paleolithic competitors. That examination of our bihemispheric brain is informed by a masterly synthesis of findings in the field of the neurological sciences by Iain McGilchrist in *The Master and His Emissary: The Divided Brain and the Making of the Western World* (2009).[3] The pioneering work of these three thought leaders places us much more confidently at the point of contact in Eurasia during the Middle Paleolithic.

THE PALEOLITHIC FOUNDRY

In *War Before Civilization,* the parallels to the Middle Paleolithic Era in Europe are compelling because Lawrence Keeley's entire focus is on the nature and role of conflict in hunter-gatherer/tribal societies prior to the influence of civilization. More to the point, the book includes anthropological research and ethnographic surveys at the scale of the "band," the small unit typical of *Homo neanderthalensis* life, which Keeley describes as consisting of *"a few related extended families who reside or move together."*

The Paiute Indians of the North American Great Basin and the Aborigines of Central Australia are cited as examples of such "band organization." Large independent and cross-cultural surveys of armed conflict in hunter-gatherer societies around the world provide a clear picture of frequent and sometimes extreme violence as a fixture in their harsh existence.

The centerpiece of Keeley's argument is the graph shown in Table 2.1 indicating the annual war fatality rates expressed as a percentage of the total population. Here he compares twenty-three primitive societies (bands, tribes and chiefdoms) from around the globe to civilized societies (Germany, Russia, France, and Japan heading the list). The primitive societies, primarily hunter-gatherers, had more than four times the number of annual casualties as a percentage of population when compared to the civilized societies.

Other sources for forensic data are the cemeteries of the Late Paleolithic, when formal burial grounds first began to provide the wide sampling of intact remains from the multiple generations that are needed to establish societal conclusions about violence and causes of death. A particularly instructive example cited by Keeley is the Late Paleolithic cemetery of Gebal Sahaba in Egyptian Nubia, 14,000–12,000 YBP:

Over 40 percent of the fifty-nine men, women and children buried in

Table 2.1. Annual Warfare Death Rates for Various Primitive and Civilized Societies

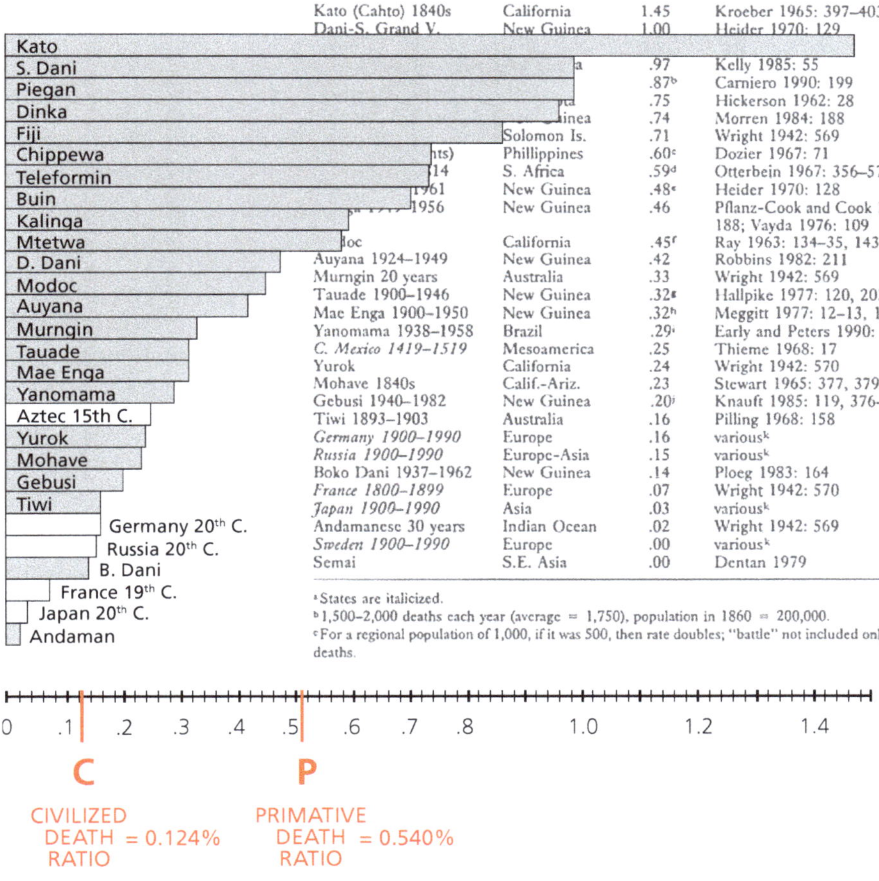
Kato
S. Dani
Piegan
Dinka
Fiji
Chippewa
Teleformin
Buin
Kalinga
Mtetwa
D. Dani
Modoc
Auyana
Murngin
Tauade
Mae Enga
Yanomama
Aztec 15th C.
Yurok
Mohave
Gebusi
Tiwi
Germany 20th C.
Russia 20th C.
B. Dani
France 19th C.
Japan 20th C.
Andaman
0
.1
.2
.3
.4
.5
.6
.7
.8
1.0
1.2
1.4
C
P
CIVILIZED DEATH RATIO = 0.124%
PRIMATIVE DEATH RATIO = 0.540%

Table 6.1 *Annual Warfare Death Rates*

Society[a]	*Region*	*Annual % Rate*	*Source*
Kato (Cahto) 1840s	California	1.45	Kroeber 1965: 397–403
Dani-S. Grand V.	New Guinea	1.00	Heider 1970: 129
		.97	Kelly 1985: 55
		.87[b]	Carniero 1990: 199
		.75	Hickerson 1962: 28
		.74	Morren 1984: 188
	Solomon Is.	.71	Wright 1942: 569
	Phillippines	.60[c]	Dozier 1967: 71
	S. Africa	.59[d]	Otterbein 1967: 356–57
	New Guinea	.48[e]	Heider 1970: 128
	New Guinea	.46	Pflanz-Cook and Cook 1983: 188; Vayda 1976: 109
	California	.45[f]	Ray 1963: 134–35, 143
Auyana 1924–1949	New Guinea	.42	Robbins 1982: 211
Murngin 20 years	Australia	.33	Wright 1942: 569
Tauade 1900–1946	New Guinea	.32[g]	Hallpike 1977: 120, 202
Mae Enga 1900–1950	New Guinea	.32[h]	Meggitt 1977: 12–13, 109
Yanomama 1938–1958	Brazil	.29[i]	Early and Peters 1990: 18
C. Mexico 1419–1519	Mesoamerica	.25	Thieme 1968: 17
Yurok	California	.24	Wright 1942: 570
Mohave 1840s	Calif.-Ariz.	.23	Stewart 1965: 377, 379
Gebusi 1940–1982	New Guinea	.20[j]	Knauft 1985: 119, 376–77
Tiwi 1893–1903	Australia	.16	Pilling 1968: 158
Germany 1900–1990	Europe	.16	various[k]
Russia 1900–1990	Europe-Asia	.15	various[k]
Boko Dani 1937–1962	New Guinea	.14	Ploeg 1983: 164
France 1800–1899	Europe	.07	Wright 1942: 570
Japan 1900–1990	Asia	.03	various[k]
Andamanese 30 years	Indian Ocean	.02	Wright 1942: 569
Sweden 1900–1990	Europe	.00	various[k]
Semai	S.E. Asia	.00	Dentan 1979

[a]States are italicized.
[b]1,500–2,000 deaths each year (average = 1,750), population in 1860 = 200,000.
[c]For a regional population of 1,000, if it was 500, then rate doubles; "battle" not included only raid deaths.

Adapted from "War Before Civilization" by Lawrence Keeley.

[The grey bars show the comparative annual death rates in the various tribes from "Kato" to "Tiwi" and the white bars show civilizations and their death rates. Keeley's supporting statistical analysis is shown in the background.]

Note: The war fatality rates of the USA during the twentieth century are too small to be visible on this chart. The U.S. Civil War, as an internal conflict, is not included.

this cemetery had stone projectile points intimately associated with or embedded in their skeletons…[H]omicidal violence at Gebel Sahaba was not a once-in-a-lifetime event, since many of the adults showed healed parry fractures of their forearm bones—a common trauma on victims of violence—and because the cemetery had obviously been used over several generations. The Gebel Sahaba burials offer graphic testimony that prehistoric hunter-gatherers could be as ruthlessly violent as any of their more recent counterparts and that prehistoric warfare continued for long periods of time.[4]

The conclusion to be drawn is quite clear: as soon as the densities of human population began to increase and the cemeteries of the Late Paleolithic began to accumulate forensic evidence spanning generations, we find evidence of a consistently brutal and violent existence at the time when *Homo sapiens* were beginning to transition upward from the animalistic baseline.

The small sampling of skeletal remains we have from the earlier Middle Paleolithic are found mainly in caves, which provide protection from the destroyers of organic remains: sun, rain, freeze-thaw cycles, and nature's cleanup crew of carnivores, vultures, and insects. By contrast, multigenerational burial sites, such as Gebel Sahaba, preserve much more: an encyclopedia of a culture's forensic data over time. The retreat behind the characterizations of "thin evidence" in the Middle Paleolithic, so often used by scientists who deny interspecies violence within the genus *Homo*, becomes much more difficult to defend. Clear evidence, physical and genetic, continues to pile up in support of a violently competitive beginning for humankind.

For those remaining in the "no interspecies violence" camp for the Middle Paleolithic period, there is the troubling fact that the most complete skeleton of *Homo neanderthalensis* (*Shanidar 3*, from Turkey) has evidence of a fatal wound, a wound that has been argued to be from a projectile weapon around the time of the *Homo sapiens* incursion.[5]

Table 2.2 provides a more detailed analysis of the frequency of violence among Native American tribes in the North American West, and reveals yet another example of the high levels of violence.

Although often referred to as warfare, the principal form of attack by hunter-gatherers was the raid, which tends to be smaller in scale and produces fewer casualties than the large-scale "civilized" conflicts, which average five-year intervals. However, because raids were so frequent, they eventually accounted for a much larger casualty rate and a far greater disruption of domestic life. Keeley's work is powerfully ordered and illustrative, an original challenge to the

Table 2.2: Frequency of Raids/ North American West

Type of Warfare	More than 4 per year	2-4 per year	None or 1 per year	Totals
Offensive Raid	44 (28%)	50 (31.9%)	63 (40.1%)	157 (100%)
Defense against Raid	52 (33.5%)	77 (49.7%)	26 (16.8%)	155 (100%)
Offensive or Defensive Warfare	68 (43.3%)	68 (43.3%)	21 (13.4%)	157 (100%)
Subtotal of all Conflict/Annum	43.3% +	43.3% =	86.6% in conflict at least 2-4 times per year	

[Adapted from War Before Civilization *by Lawrence Keeley. Source: Joseph G. Jorgensen,* Western Indians: Comparative Environments, Languages, and Cultures of 172 Western American Indian Tribes *(1980)].*[6]

myth of the peaceful savage. A much more comprehensive and updated reference expanding Keeley's work and that of many others is Steven Pinker's *The Better Angels of Our Nature.*[7]

Occasionally, particularly at the larger scale of the tribe, massacres occurred, usually carried out by one community or tribe with the intent to annihilate others. A particularly instructive example is drawn by Keeley from the supposedly idyllic period of Native North Americans prior to the arrival of Columbus. In an untouched land of plentiful game, we would expect to encounter the often-heralded "noble savage," but we find just the opposite.

Crow Creek was the site of just such a large-scale massacre. The village layout, with surrounding fortifications and cliff faces, is described in Figure 2.1, and Keeley gives a detailed summary of findings:

At Crow Creek in South Dakota, archaeologists found a mass grave containing the remains of more than 500 men, women, and children who had been slaughtered, scalped, and mutilated during an attack on their village 150 years before Columbus's arrival (ca. A.D. 1325). The attack apparently occurred just when the village's fortifications were being rebuilt. All the houses were burned, and most of the inhabitants were murdered. This death toll represented more than 60% of the village's population, estimated from the number of houses to have been about 800…The survivors appear to have been primarily young women, as their skeletons are under-represented among the bones; if so, they were probably taken away as captives. Certainly, the site was deserted for some time after the attack because the bodies evidently remained exposed to scavenging animals for a few weeks before burial. In other words, this whole village was annihilated in a single attack and never reoccupied.[8]

Phillip L. Walker, in his paper "A Bio-archaeological Perspective on the History of Violence," elaborates on Crow Creek:

The conclusion that this massacre was the result of inter-village warfare is reinforced by on-going research that has produced evidence of similar massacres at two fourteenth century villages within striking range of Crow Creek.[9]

Walker also addresses the broader canvas of violent competition and interpersonal violence as an inherent feature of the human condition over time:

Bioarchaeological research shows that throughout the history of our species, interpersonal violence, especially among men, has been prevalent. Cannibalism (in Prehistory) seems to have been widespread, and mass killings, homicides and assault injuries are also well documented in both the Old and New Worlds. No form of social organization, mode of production, or environmental setting appears to have remained free from interpersonal violence for long.[10]

Keeley also underlines the universality of the Native American findings:

After surveying a large number of prehistoric burial populations in the eastern United States, archaeologist George Milner concluded that the pre-Columbian warfare of this whole region featured "repeated ambushes punctuated by devastating attacks at particularly opportune moments." From North America at least, archaeological evidence reveals precisely the same pattern recorded ethnographically for the tribal peoples the world over of frequent deadly raids and occasional horrific massacres.[11]

Drawing on this background, including the forensic and archaeological evidence cited earlier, there are three baseline characteristics of the hunter-gatherer bands

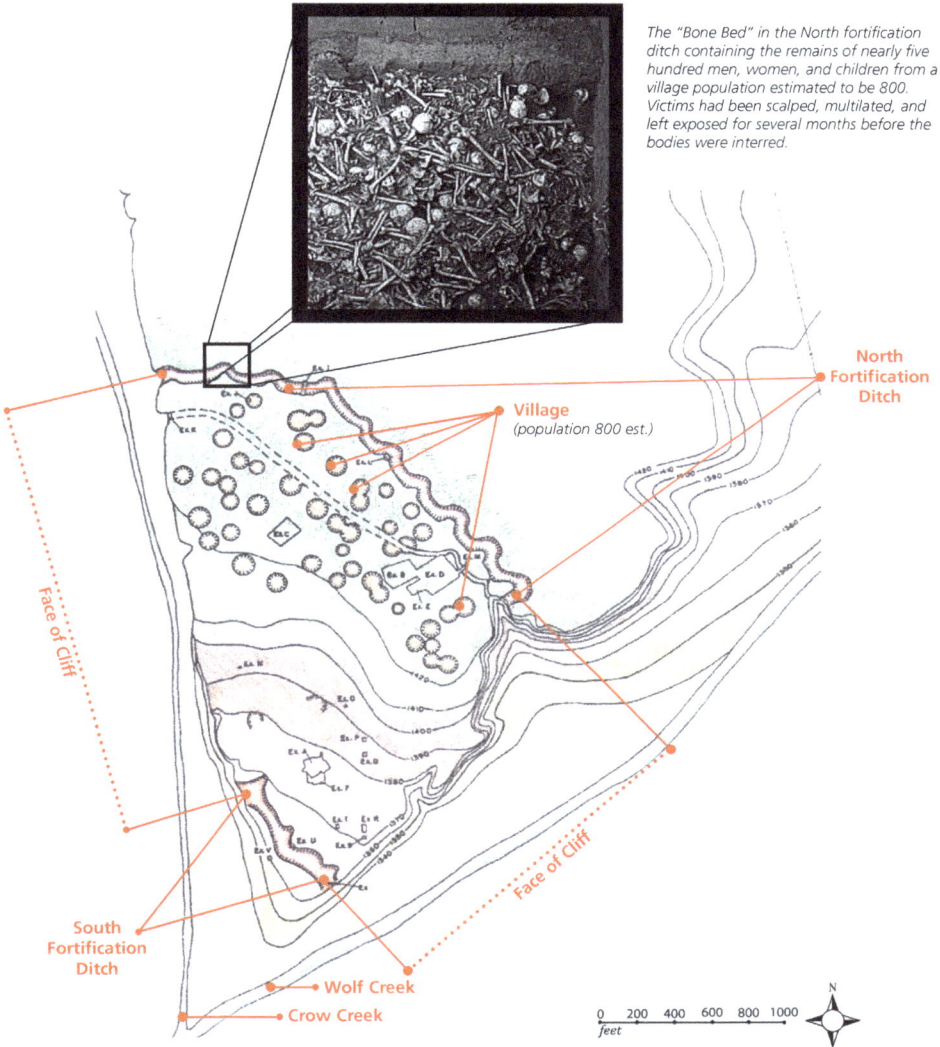

The "Bone Bed" in the North fortification ditch containing the remains of nearly five hundred men, women, and children from a village population estimated to be 800. Victims had been scalped, mutilated, and left exposed for several months before the bodies were interred.

Figure 2.1. Pre-Columbian Massacre at Crow Creek, South Dakota (690 YBP)

Adapted from War Before Civilization by Lawrence Keeley. 1980

of *Homo neanderthalensis* society in Eurasia, prior to the arrival of *Homo sapiens* about 42,000 YBP, which we adopt herein as working assumptions. We will later support these assumptions as the precondition for the pattern of early death before reproduction that shaped our uniquely human natures.

Baseline Characteristics

- *Homo neanderthalensis* lived in small bands (5–8 in each band, possibly 24–28 in a temporary grouping of bands for defense) located near a cave or rock-sheltered area, and typically near animal migration routes and water. A high level of competition would have existed for the deep cave shelters, which never go below freezing even in the harshest winter, as well as for the adjoining hunting grounds, and for the most desirable mates. Conflicts as acts of revenge for past deeds or other issues of dominance and control would be common.

- Inter-group conflict (consistent with primitive human societies) would include raiding of other bands, maintaining a strong defense and, in both instances committing murder, rape, and torture.

- Representing an *"über-*indigenous" people who had achieved a 300,000-year presence and 200,000-year (10,000-generation) dominance in Eurasia—prevailing through Ice Ages and volcanic winters and never hesitating to kill prey or foe—*Homo neanderthalensis* was the embodiment of fierce territoriality and the unflinching will required for survival. They would not go willingly from their verdant land, defensible topography, and life-saving caves for winter shelter. They would fight.

THE ADVANTAGED AGGRESSOR

Keeley, Walker, Milner and their archaeological colleagues have given us, at the highest levels of probability, a universal model of the violent context within which the hunter-gatherers of the Middle Paleolithic would have competed for survival. To this author, these immediate parallels leave no doubt that both *Homo neanderthalensis* and *Homo sapiens* were fully practiced in bloody confrontations on offense and defense long before they met. Given that the staggered prehistoric arrival of *Homo sapiens* was approximately 42,000–40,000 YBP (first in the Caucuses and later into Central Europe); can we find a similar incident that is grounded in this geographical context? Is there a close cultural parallel to the confrontation and replacement that took place?

The answer is yes to both questions. While we in the Americas are most familiar with the fall of New World indigenous civilizations, conquered by the European powers of Spain, England, France, and Portugal in the fifteenth and sixteenth centuries, it is the less familiar takeover of Central Europe by the vanguard of the Neolithic Revolution approximately 8,000–6,000 YBP that provides the most powerful parallel, the "mirror event" of the Middle Paleolithic.

In the Neolithic Revolution we see the familiar formula of a mobile society in possession of military advantage and acting in the role of aggressor. Jared Diamond's book *Guns, Germs, and Steel* is specifically relevant to this time frame and these events.

Diamond, a professor of geology and physiology at UCLA, addresses the big picture of large-scale variations in geographical, flora/fauna, and agricultural resources as they drive various societal development trajectories. But it is his observations on the pattern of repeated violent collisions between advantaged aggressor societies and indigenous peoples that has such relevance to our considerations of the Middle Paleolithic:

> The history of interactions among disparate peoples is what shaped the modern world through conquest, epidemics and genocide. Those collisions created reverberations that have still not died down after many centuries, and that are actively continuing in some of the world's most troubled areas…Much of human history has consisted of unequal conflicts between the haves and the have-nots.[12]

The Neolithic Revolution took place on the same lands contended for by *Homo neanderthalensis* and *Homo sapiens* in the defining conflict of the genus *Homo*, but this time, the hunter-gatherer victors of that early conflict were in the indigenous/disadvantaged role. Their fate was to be violently displaced by the more powerful agrarian culture, the *Linearbandkeramik* (which translates as Linear Band Pottery Culture and is referred to herein as LBK). They followed a very similar pathway, migrating generally from Turkey and the Near East as they moved northwest into Central Europe and took possession. Jared Diamond observed:

> Twelve thousand years ago, everybody on earth was a hunter-gatherer; now almost all of us are farmers or else are fed by farmers. The spread of farming from those few sites of origin usually did not occur as a result of the hunter-gatherers' elsewhere adopting farming; hunter-gatherers tend to be conservative…[I]nstead, farming spread mainly through farmers' outbreeding hunters, developing more potent technology, and then killing the hunters or driving them off of all lands suitable for agriculture.

The key competitive component in the ascendancy of one human culture over another, the breeding of a larger population, depended not only on how much land could be controlled but also on how much food value (calories) could be extracted per unit of land. The more efficient method will create the most abundant and varied harvest, resulting in the greater population density and thereby the greater number of young males who will be available for combat. This was the essence of the agricultural "engine" that eventually drove the massive scale of the Egyptian civilization. A three-stage simplification of these revolutionary breakthroughs follows:

> **Stage One:** Foraging for wild edibles, stalking and killing primarily large animals with thrusting spears. This method of *Homo neanderthalensis* has been estimated to require a vast amount of land and therefore to have a population-limiting effect within the area of control.
>
> **Stage Two:** Expanding the group, social structure, and scope of the hunt to include fast and elusive small and medium-sized mammals, birds, and fish through the use of snares, nets, and projectile weapons. This advanced method of the *Homo sapiens* supported, over time, much larger population densities, and ultimately they appear to have out-bred their fellow members of the genus *Homo* and employed superior technology in their elimination.
>
> **Stage Three:** The ultimate ultimate food value strategy is agriculture plus hunting and the domestication of animals. This supported the yet-larger densities of the Neolithic *Homo sapiens* farmers (LBK) in their conquest of the Mesolithic *Homo sapiens* hunter-gatherers who survived the Last Great Ice Age only to be erased in their European homelands.[13]

In 2007, Mark Golitko and Lawrence H. Keeley authored *Beating Ploughshares Back into Swords: Warfare in the "Linearbandkeramik,"*[14] which brings this violent transition from hunter-gatherer to agrarian settlement into clear focus. Although the Neolithic Revolution was a relatively uneven process, the LBK expansion into Central Europe provides a prototypical expression of this turning point in human history. Golitko and Keeley built on a knowledge base of "the best-studied Neolithic culture in all of Europe, with hundreds of sites having been subjected to excavation over the last century,"[15] so they were not hampered in telling their story by the meager remains that charcterize the Middle Paleolithic.

These sites provide explicit witness, in the form of settlement fortifications, massacre locations, and a wealth of skeletal remains, to the violent nature of the replacement of the hunter-gatherers by the agrarian-potter culture. The continuing severe and competitive violence between factions of the victorious LBK long after the defeat of the Mesolithic hunter-gatherers reaffirms the universal nature of this violence.

Various "peaceful" counter-arguments in the scientific community have suggested a slow dispersion of the LBK into Central Europe with an extended period of trade, accommodation, and cooperation as the LBK began to integrate with, and eventually absorb, the Mesolithic hunter-gatherers. Golitko and Keeley however, cite current archeological data that supports an inward migration by the "sudden appearance of a radically new material culture [linear band pottery] and subsistence system [agriculture]." The scientists point to the most recent radiocarbon dating techniques that support a Mesolithic/Neolithic transition in the LBK region that would have been "quite abrupt."

Further confirmation comes from Peter Rowley-Conwy in the October 2011 issue of *Current Anthropology*. Citing advancement in excavations and dating methods across Europe, Rowley-Conwy concurs:

> The LBK remains a sharp archaeological event, still best interpreted as a [invasive] migration.[16]

Close archeological examination of these well-preserved settlements from the Neolithic takeover of Central Europe by the LBK confirms their militarily sound planning and their close attention to water, defensive posture, and the "gold" these agriculturalists sought: the uniquely productive Loess soils of north-eastern Europe.

The later stages of replacement in the west most clearly demonstrate the military mindset and strategic thinking of the LBK. Figure 2.2 maps out a specific example of these opposing groups. In positioning their two major settlements, to control the valuable loess soils, they chose the east and west sides of the Meuse River Valley, a highly defensible topographic cleft that is now the boundary between Belgium and the Netherlands.

The dramatic landforms and sheltering character of the Meuse River Valley offer a parallel to the river valleys of the Dordogne and the Ardeche in southern France, the highly defensible positions for *Homo neanderthalensis* in their final days. These topographic features were the natural castles and citadels of their day and were readily exploited for their military advantage. The strategic exploitation of topographic features is also seen in the cliff dwellings of Mesa Verde in Arizona in the American West. There the enemy is forced by rock formations into an easily defended single-file approach, a strategy that we see employed in the fortifications of the Neolithic Period.

The typical Neolithic Defensive Perimeter, as shown in Figure 2.3, employed a surrounding palisade wall of tree trunks protected by a fortification ditch. Only a narrow opening in the ditch allowed access to the portal of entry, which forces aggressors into a most vulnerable position—the perpendicular single file. The ditches are precursors of the familiar moats and elaborate surrounding trenches of the Middle Ages, and the "gateways" in the palisade wall eventually became drawbridges. From an upper walkway built into the palisade wall, the defending archers could shoot down on attackers as they struggled up the hills that typically surrounded such positions.

This is an advancement of the high ground and high visibility defensive locations that were chosen by *Homo sapiens* in their earlier takeover of Europe from *Homo neanderthalensis*. The effectiveness of the fortification ditches that funnel the attackers into the killing zone is easily readable by the mass of arrowheads discovered at these gateways.

Fortifications and forensic evidence of traumatic death greatly increased in this later period of the western expansion, which eventually involved an extended and virulent period of conflict between LBK settlements. After the violent expulsion of the hunter-gatherers, they started attacking each other! The frequency and scale of violence in this later period approached crisis proportions "equaling the highest sustained levels of injury and lethality of conflict ever recorded in even the most violent of civilizations."[17] This is not behavior that just arises with the advent of an agrarian society; this is the continuation of an unbroken advancement in the ways and means of brutal conflict, a biological imperative in the competition for survival and domination.

A shared characteristic of several of the sites of LBK massacre—an early example being Schletz-Asparn near Vienna, Austria, and a late example being Talheim in Germany's Rhine Valley—is that there was an underrepresentation of women of childbearing age. In Europe today, there is a large Neolithic genetic representation across the population, as well as a much more homogeneous genetic profile among women compared to men. Collectively, this is strong evidence of LBK genetic domination.

This supports the aggressive "lunging advance of LBK and genetic mixing," Peter Rowley-Conwy points out. He goes on to state: "The way to destabilize and Neolithize [genetically overpower] hunter gatherers…is to encroach on their territory and steal their women."[18]

This highly readable genetic imprint, which results from victors breeding with the females of the vanquished or oppressed, occurs throughout history but is particularly notable in the massive genetic imprint (present in 8 percent of the male population in conquered lands) of Genghis Khan in the Middle Ages.[19]

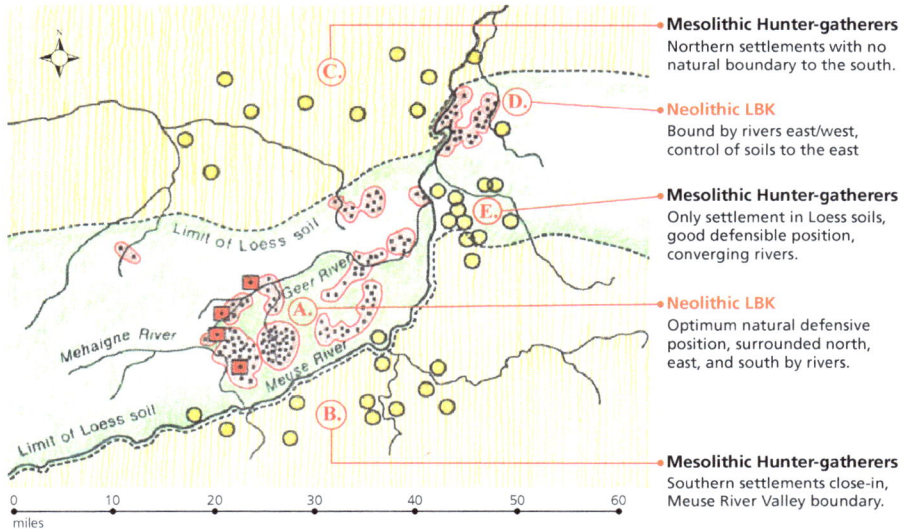

Figure 2.2. LBK Strategic Settlement: Meuse River Valley

This river valley was the western boundary of the Holy Roman Empire in 1301, and it was through this natural fortress that the Germans launched their invasion of France on August 14, 1914. The advantages of the LBK's strategic planning are clearly visible in the placement of their largest settlement (A), so as to be surrounded on three sides by rivers with four forts protecting the exposed western end. The second largest settlement further north (D) is positioned between two converging rivers, gaining control of the fertile loess soils to the east (loess soils appear in green).

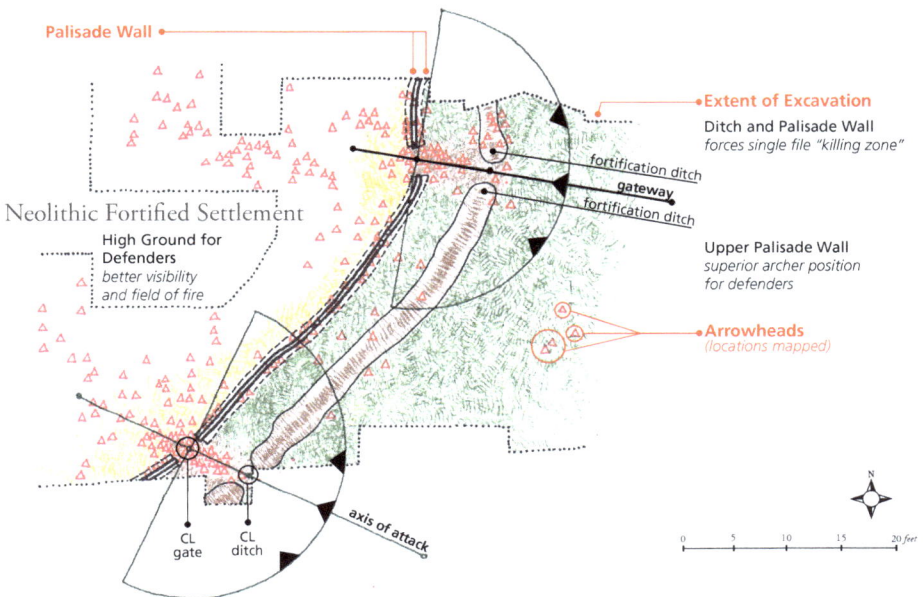

Figure 2.3. Neolithic Settlement Defensive Palisade Perimeter

Adapted from War Before Civilization *by Lawrence Keeley, 1980 (Figs 2.2 and 2.3).*

In the annals of his conquests Genghis Khan was reported to have said:

> The greatest happiness is to scatter your enemy, to drive him before you, to see his cities reduced to ashes, to see those who love him shrouded in tears, and to gather into your bosom his wives and daughters.[20]

In a pride of lions that has been taken over by a dominant male, there is the similarly motivated killing of all the cubs sired by the displaced male.[21] This seemingly impossible role thrust upon female lions has a biological mirror in female mice who spontaneously reabsorb fetuses upon the emergence of a newly dominant male (the Bruce effect)[22] and also in geladas, relatives of baboons, who exhibit "...a wave of spontaneous abortions"[23] in such an event.

Female chimpanzees who typically must migrate from their home group (incest avoidance) and gain acceptance within the often hostile setting of other groups suffer no more than mild discomfort compared to this violent pattern. Our female ancestors faced the same terror in that they were abducted by, and faced sexual accommodation or death at the hands of, the males who had just killed their families.

A powerful telling of such a story is S. C. Gwynne's book *Empire of the Summer Moon*, the story of Cynthia Ann Parker, a young white girl on the Texas frontier, who lived through such a violent kidnapping by Comanche at the age of nine, who eventually gave birth to Quanah Parker, the last chief of the Comanche tribe, and who repeatedly refused opportunities to leave the Comanche way of life. It is hard to imagine that such instances are not biologically related to what we refer to as the Stockholm syndrome, which is expressed as "capture-bonding"—empathy and identification with captors, which occurs primarily in females.

Such patterns represent a broadly based and violent dynamic in nature. We may consider such events to be exceedingly rare for the females of our species, but this phenomenon has had a powerful evolutionary impact. In our lifetimes we have witnessed the reenactment of this insult to humanity by hundreds if not thousands of males in Rwanda, Darfur, Sudan and, most recently, by Boko Haran and ISIL (a.k.a. ISIS).

INTERBREEDING: SEXUAL ATTRACTION AND REPULSION

For all the violence and oppression that attended the Spanish arrival in South America (land of the Aztec, Maya, and Inca), the European arrival in North America (land of Native North American Indians) or the British arrival in Australia (land of the Aborigines), direct descendants of all these native peoples have survived into modern times. This is certainly not the case with *Homo neanderthalensis*, who is extinct but whose minor genetic legacy of approximately 2.0 percent of our DNA suggests that interbreeding, if biologically viable, may not have been culturally or sexually desirable.

How could *Homo neanderthalensis* be effectively eliminated from the face of the earth when they had established a vast European presence as a successful and self-sufficient culture for over two hundred thousand years? Were they too different, too alienated genetically and physically, when *Homo sapiens* arrived 42,000 YBP? To answer this question, we need to consider the elemental aspects of reproduction and the related emotions of human sexual attraction and repulsion.

Job one in evolution is the replication of life forms—not approximations, but reproductions that are highly accurate and therefore viable. In the earliest forms of life, the reproduction method was "asexual," no genders or sexual partners required, just the direct reproduction of the life form. This method is still in use, from strawberries to hammerhead sharks. However, it has drawbacks when compared to sexual reproduction which has the advantage of fewer errors or "misprints" reaching maturity and projecting into the next generation.

Many other advantages such as higher diversity and complexity also attend sexual reproduction but it does come with a prohibition: avoid incest, that is to say, sex between family members. If close kin interbreed, they are more likely to bear children affected by recessive and deleterious traits of the parents. For this reason, among all the inclinations we inherit, the aversion to incest is one of the most powerful. The near universal sexual aversion between siblings is known as the "Westermarck Effect."[24]

Being sexually attracted or repelled depends primarily on our facility of pattern recognition. The central role of this feature, evident to scientists for many years, has become more clearly understood recently through neuroimaging technology that provides a more detailed understanding of the geography of the brain (what goes on where) and the dynamic interaction of these capabilities (what goes on when) in our conscious and unconscious mind. The large amount of the brain that scientists observe as interconnected with pattern recognition confirms the high criticality of this function, literally one of life or death, as well as its further role in our selection of evolutionary and reproductive pathways.

Amazingly, while our planet currently has a population of over 7 billion people, almost all of whom have two eyes, a nose, and a mouth of approximately the same proportion and location, we can—with almost perfect accuracy—distinguish between each of them, even in the case of "identical" twins. That our skill in pattern recognition can defeat nature's best attempt at genetically identical humans is the ultimate confirmation of its high level of resolution and, therefore, importance. This almost superhuman ability to instantaneously evaluate even the slightest of variations in the visual presentation of another being is about as necessary as a survival skill ever gets.

For early humans, this ability created a heightened survivability through early recognition of an enemy, a predator, or prey. Failure to distinguish between an approaching friend and an enemy in the shadows of late evening or to sense the subtle shading of a cave lion crouched silently within heavy vegetation could be fatal. Those without a highly developed sense of pattern recognition and the ability to read causation into the most subtle of environmental changes would quickly be on the list of extinct species. This attribute was under constant pressure for advancement in *Homo sapiens* as a result of surrounding and upward-evolving competitors in the wild and in competing members of the genus *Homo*.

No less important is the role of pattern recognition in the discernment of beauty, the physical ideal of each culture to which one is irresistibly drawn. Sexual attraction, selection through competition, mating, and reproduction is Survival 101 and is deeply informed over time by the group's evolving culture and unique natural environment. Separated groups of the genus *Homo*, as generations pass, will follow their own path of natural selection and acquire an increasingly distinct physical and cultural identity; an unshakable consensus: "us" as opposed to "them." The corollary is also true: in evolving a heightened sense of their ideal or norm, early humans created an internal group basis for what is abnormal, disfigured, or ugly.

We still experience this negative side of our discernment when we are repulsed or shaken by the initial sight of disfigurement in a newborn child. Of course, a discernment of variance from the species norm by a mother cat or bird will quickly result in eviction from the litter or nest: a death sentence on the spot. This brutal form of self-editing for conformity within animal groups in the wild is an extension of the evolutionary process that favors a highly developed and ideal match to the environmental/biological context, and is a self-correcting aspect of sexual versus asexual reproduction.

Even at what we consider to be one of civilization's high points, under The Twelve Tables of Roman law, a child born with a visible deformity was required to be killed.[25] This underlines one of nature's primary objectives, the avoidance of variance and mutations that reduce the viability of offspring.

Is pattern recognition also part of an integrated neurobiological filter (including the Westermarck effect) that acts in concert to enhance the net effectiveness of sexual reproduction and thereby assure the highest correlation to the species norm? Isn't the persistent and reflexive nature of xenophobia and racism an echo of just such a primal reproductive force, and isn't it a sure marker of civilization when we recognize these as paranoia and inequity and set them aside?

Sexual attraction and repulsion function to encourage potentially healthy diversity or hybridization within closely grouped branches of a species or, in the case of a more emphatic break across the species boundary, to discourage a genetically negative pairing that would not be viable. This raises the question of what differences had cropped up in those 458,000 years that had passed since these two competitors diverged from a common ancestor. At the point of first contact, was there an issue of biological incompatibility?

For all the similar human qualities that *Homo neanderthalensis* embodied, they would still have looked dramatically different, a close impersonation of Mary Shelley's *Frankenstein*, brutish, powerful and by our view, ugly. Squat in comparison, with disproportionately large eyes, heavy brow, and a rounded "bun" at the rear of their skull, they would not have appeared to be fully human.

Our constant fascination with mutants and near-human assemblages— *The Island of Dr. Moreau* comes to mind—continues to be a central thematic element of modern horror literature and film. The Middle Paleolithic reality of near-human competitors lurking nearby is seemingly imprinted on our brains because overzealous witnesses continue to see, or claim to see, Sasquatch or Big Foot in that dark, deep forest just beyond the edge of our knowledge, a nightmare of a large half human, half ape that apparently refuses to leave our collective consciousness.

A healthy regard or fear of *Homo neanderthalensis* would have come from knowing that they could silently stalk an objective, in low-light conditions, with a fire-hardened spear and, emerging ghostlike out of the night, gut a deer or the vulnerable mid-section of *Homo sapiens*. While these are hunting and tracking skills similar to those of Native Americans, with an extra 140,000 years to perfect their craft, *Homo neanderthalensis* may have been the ultimate native warrior.

Beyond body type, what would have been the greatest, most shocking difference between these two (still under debate but well-supported in their genetic profile)? *Homo neanderthalensis* had white skin! *Homo sapiens* and all their African ancestors to that date were black-skinned, with hair to match. They must have considered it a very strange sight, all those stocky, pale, even ghostly, bodies with a rainbow of hair colors!

On the other hand, upon seeing essentially Ethiopian body types—tall, thin and gracile in comparison—*Homo neanderthalensis* would have been less shocked. Their ancestors were black, they still had babies arriving in ranges of light to darker skin, and they had confronted and destroyed the occasional black-skinned, odd-looking near-human intruder. Eurasia was their turf, and although they had cleared out all competitors hundreds of thousands of years before, there was still the occasional pretender to top hominid knocking on the door.

Depigmentation, a process of transitioning from black skin to white skin over tens of thousands of years, is an adaptation that aided in the processing of vitamin D at the less sun-drenched northern latitudes of Europe versus Africa;[26] the transition has been genetically projected to require 10,000–20,000 years.

Although there is debate concerning the speed at which such a change takes place, we must consider the very real possibility that *Homo sapiens* carried out the extermination of the first white race of the human family: *Homo neanderthalensis*. Over time at that northern latitude, *Homo sapiens* eventually acquired white skin color.

By the time *Homo sapiens* approached the Eurasian territories, *Homo neanderthalensis* had evolved an alien physical appearance. Moreover, the genetic and sexual gateway to interbreeding that would have existed between black-skinned cousins tens of thousands of years before in the Levant (Manot Cave of present-day Israel) was either closed or closing. While we know that interbreeding took place and that we carry a small genetic remnant as confirmation of this fact, no living member of *Homo neanderthalensis* walks the Earth.

ANIMALISTIC CONFLICT VERSUS "CIVILIZED" CONFLICT

What are we to conclude when an advantaged aggressor, the African hunter-gatherer *Homo sapiens*, takes over the lands of the European hunter-gatherer *Homo neanderthalensis* 40,000 YBP, pushing them to extinction, and yet we display only a genetic pittance in common as a result? By contrast, the incursion 32,000 years later of a newly advantaged aggressor, the Near Eastern LBK agrarian/settlers, taking over the same lands, had a massive genetic impact that has recently been confirmed to be the dominant ancestral genetic expression in the population of Europe today.[27]

One explanation would be that these were collisions between fundamentally different human ancestors. In the Middle Paleolithic, it was an interspecies conflict between the most highly evolved members of the genus *Homo*. All members of the genus had weapons and the use of fire, and all had advanced, to some degree, toward self-awareness. The projection of these skills outward to the external environment confirms this commonality, but the differences remained significant. *Homo sapiens* and *Homo neanderthalensis* looked at each other across a chasm of difference, a near-animalistic human observing an almost anatomically and neurologically modern human. In fact, they were so different as to be "on the edge of biological incompatibility."[28]

The full significance of these differences has only recently come to light through the genetic analysis of high-resolution DNA from *Homo neanderthalensis* (HN) remains and subsequent comparison to *Homo sapiens* (HS) in the following papers:

1. The offspring of HS/HN mating are identified as male-hybrid sterile, which helps to explain why we have only about a 2 percent genetic contribution from HN. Sriram Sankararaman, David Reich, and Svante Pääbo and their team published this pivotal finding in "The genomic landscape of Neanderthal ancestry in present-day humans" in the online publication of *Nature* (January 29, 2014).

2. Three months later, Pääbo and his team published "Patterns of coding variation in the complete exomes of three Neanderthals" in the online *Proceedings of the National Academy of Sciences* (April 21, 2014). This study confirmed that genes involved in behavior had changed less (were less complex) in the HN lineage compared to HS. Further genetic findings confirm that the genetic diversity of HN was less than HS and indicate that HN populations were small and isolated from one another.

This Middle Paleolithic confrontation, 42,000–40,000 YBP, which extended over several thousand years, heralded, in my opinion, the emergence of *Homo sapiens* as the one life form to occupy the ecological niche of full self-awareness. The biological distance separating them from all other members of the genus was so great that although there was some interbreeding, it was characterized as being of "low magnitude" by Pääbo and Reich in their comprehensive paper of December 2013, "The complete genome sequence of a Neanderthal from the Altai Mountains."[29]

The confrontation at the top of the genus *Homo* followed a particularly agressive pattern in which *Homo sapiens* prevailed and ultimately eliminated all others. There was essentially no major gene flow due to a significant measure of biological and, in all probability, sexual and cultural incompatibility.

The Neolithic confrontation, on the other hand, was the new model of *Homo sapiens* versus *Homo sapiens*, where there is both biological and sexual compatibility. This model, in the worst case for the defeated, ended with men and boys of all ages dead and women and girls of all ages dead, with the exception of young women of childbearing age, who were raped and, if not killed, abducted.

The victorious males were the initiators of the significant gene flow that is dominant across modern generations of the European population. No more dramatic confirmation of this phenomenon can be found than the genetic analysis of European ancestry authored by David Reich and team in "Ancient human genomes suggest three ancestral populations for present-day Europeans," published in *Nature* (September 18, 2014).[30] The LBK are confirmed to be, by far, the largest genetic component of Europeans, confirming the "traditional" behavior of the victorious, still observed in the ISIS and Boko Haram male-dominant aggressors of our day.

PALEOLITHIC MINDSETS

There are two great instructors on the nature and formation of our mindsets. The first is ancient: the two very different patterns of early death experienced by prehuman males and females of our species dating back to our animalistic beginnings and still observed in our closest living relatives, the chimpanzees. Natural selection, over time, distilled the population of minds on each pathway, culling all but the most viable, those possessing the best strategies and dispositions for survival in their respective worlds. Taken together, the two very different worldviews of male and female represent the collective knowledge, the two "natures" of the human species.

The second instructor is a modern one, the cutting-edge insights achieved in the sciences of neurobiology and their mapping of the characteristics of the human brain. Here we begin to see the way in which ancestral legacies have been organized in our brains and the neural pathways by which we access, consider, and often debate the alternatives that "come to mind."

We begin with a detailed consideration of how the male and female circumstances of risk and early death were different and what qualities would have saved them into our respective ancestral family trees. While the process of amplifying those advantageous characteristics through natural selection had been happening in the surviving minds of all members of the genus *Homo*, it clearly happened differently in the pathway to the self-aware mind of *Homo sapiens*.

Natural selection is driven by early death—that is, death before reproduction—so we begin there, with Table 2.3. In the category "Worldview/Bias," we see opposite outcomes for males and females expressing empathy, trust, and flexibility. Males were in the unforgiving frontline of conflict where hesitation in an ambiguous situation or concern for the enemy can result in immediate death. Females migrated from their home group and had to insert themselves in a new group and successfully mate, preferably with the strongest male for protection. Only through flexibility and high emotional intelligence could she achieve acceptance within the new group for herself and her offspring. And yet this was not the highest moment of risk for the female. As we have seen, many females of childbearing age are abducted in raids and massacres and forced to accommodate sexual assaults. Those who survived this ordeal had to realign their loyalty to the victorious group (or convince the abductors that this is the case).

Table 2.3. Male and Female Formative Forces:
Early Death/Natural Selection

I. Prehuman Male: Prototypical Early Death

Role	"Point-of-the-Spear" aggressor or defender to the death in order to maintain the secure perimeter necessary for the extended rearing of children: must detect and defeat all external threats.
Life & Death Moments	1. Defend against attacks, raids or encroachment by "Others" on territory and, in reverse, raid and kill others. 2. Unexpected encounter with "Other". 3. Member of "Other," wounded, seeks mercy or shelter.
World View/ Bias	**No Trust / Inflexible / Not Empathetic** Those exhibiting trust or hesitation in life-or-death moments would, in most cases, have died an early death. Natural selection, over time, would perpetuate aggressive first-reaction, instant-response to protect status quo. (Males remain with their home group for life).

II. Prehuman Female: Prototypical Early Death

Role	Continuity/survivability of genetic lineage, and nurture of next generation.
Life & Death Moments	1. Overcoming opposition and conflict to migrate (avoid incest) and join new family group. Select/achieve mating with top-rank male in new group for best protection. 2. Surviving kidnapping, assault, rape; building rapport in hostile settings, achieving acceptance; flexible, possessing humor, creative. 3. Reconciliation of continuing opposition and conflict with females of new group.
World View/ Bias	**Trust/Flexible/Empathetic** Mission requires migration to new groups, emotional intelligence, trust, and commitment to new community, as well as adaptability to change and to new environments. Those exhibiting aggression and resistance (particularly in instances of capture by enemy) would be killed. Skills to reconcile or diffuse aggression and to gain acceptance and security for offspring were required for survival.

The harsh reality was that the majority of males expressing trust and empathy were killed and the majority of females expressing outward aggression and resistance were killed.

As a result of such patterns, natural selection, working through hundreds of thousands of generations deep into our unbroken chain of life—at least 7.5 million years—selected for two very different versions of a viable human life experience, both of which have been recorded in our DNA. However, for reasons we will later examine, involving the role of RNA in the formation of the bihemispheric brain, both life experiences inform our mindsets and we do not turn out to be as totally different as our anatomical gender would suggest.

MAPPING THE BIHEMISPHERIC BRAIN

The science of neurobiology and the tools of neuroimaging, such as functional magnetic resonance imaging (fMRI), positron-emission tomography (PET), and magnetoencephalography (MEG), now give us a more holistic view of our bihemispheric brain in operation. My assertion is that the distribution of "worldviews" revealed in these mappings of the brain reflects the male-lineage and female-lineage characteristics we observed in Table 2.3.

While there are many books and papers on the subject, Iain McGilchrist has written what I find to be a uniquely integrative and insightful interpretation of these findings. *The Master and His Emissary* expertly explores the deeper implications of the differences and inter-relationship of the hemispheres of the brain.[31]

While McGilchrist is quick to point out that he does not attribute gender influences to the division of the brain, his astute characterizations are most relevant to our considerations here. These writings are examined in greater detail in Chapter Four, but here they provide basic guidance for our Middle Paleolithic overview, a baseline summary:

- Our brains are not only divided in a literal, anatomical sense, but each hemisphere has its own view of the world, its own "take" or predisposition.

- Certain skill sets or points of view may be located in one or both of the hemispheres. When located on both sides, they are not separated, since a complex set of interconnections provides the avenues through which a deliberation occurs between the two hemispheres.

- We frequently experience the neurological phenomenon of interconnectedness as an internal debate between worldviews as we consider an action to be taken or a conclusion to be drawn.

McGilchrist maintains that the right hemisphere (the Master) embodies the higher values of civilization, while the left hemisphere (the Emissary) tends to be aggressive and inflexible and is currently dominant and becoming more so in the Western World. Much of McGilchrist's book is an exploration of the supporting evidence and future implications of this shift in the modern world. Such insights into the brain's hemispheres and their struggle for dominance provide a uniquely productive model for understanding the mindsets of *Homo neanderthalensis* and *Homo sapiens.*

McGilchrist builds his argument on a wide range of findings in the neurosciences and draws a compelling and detailed portrait of the nature of each of our brain's hemispheres, which I, most certainly, will not undertake herein. Instead, I have interpreted from his descriptions a baseline of characteristics that are central to these Middle Paleolithic mindsets. The universal characteristics that appear in Table 2.4 would have been present in the brains of both adversaries but, as we will see, the greater influence of the right hemisphere proved to be decisive for *Homo sapiens.*

By a process of reverse engineering, we can read the unique nature of these two ancestral minds from what we have come to know about their skill sets, artifacts, cultural expression, weapons, and modes of combat. Understanding what occurred between these competitors 42,000-40,000 YBP is a process of self-examination since the fundamental organization and nature of our "two minds," as shown in Table 2.4, is unchanged from then to now.

From the beginning, the personality of *Homo neanderthalensis* was clearly expressed in their settlement patterns, which typically would be close to the migratory routes of animals and adjoining a stream or river. This puts one in mind of the answer attributed to Willie Sutton when asked why he robbed banks…; "Because that's where the money is." By picking an easy solution to food and water, they often ended up at a strategic disadvantage, since low-lying terrain is inherently more difficult to defend. Even an existing cave at higher elevations is more difficult to defend than a fortified position on high ground with visibility to all approaches.

Most revealing is the low level of cultural advancement by *Homo neanderthalensis* over 200,000 years. Weapons, the use of fire, clothing, and shelter, were only marginally advanced over more than 10,000 generations. The early physical dominance of *Homo neanderthalensis* against all challengers, assured by stealth, superior night vision, and brute strength, followed by an extended period of no challenges, led to a mindset that had not been sharpened by new threats.

Theirs is a classic example of maintaining the status quo while becoming more inflexible of method and mind. While we have increasing evidence of their burial of the dead, and use of seashells for jewelry, and pigments for body paint, none of these expressions ever approached the creative level of the art, music, sculpture, or more elaborate burial rituals that characterized *Homo sapiens* at the time of their confrontation.

As a decision of the first order and based on their settlement patterns, *Homo sapiens* concluded that high ground and visibility in all directions were the essential elements of a defensible position. Since they exhibited a habitual and successful pattern of expansion and incursion into other lands, it is not unusual that they anticipated attack from their hyper territorial peers. While these more defensible locations meant that *Homo sapiens* would have to travel further for food and, in order to survive the cold in such exposed locations would have to compensate through more labor-intensive strategies (clothing and constructed shelter), they knew that living to fight another day came first. This anticipatory preparation and vigilance was a matter of life or death in the Paleolithic, Mesolithic, and Neolithic eras. Unlike *Homo neanderthalensis*, who mainly occupied caves for shelter, *Homo sapiens* took advantage of the stable temperatures and the absence of a freeze-thaw cycle to create underground galleries for their art, rituals, and ceremony.

Reading the land to exploit its advantages would have been a key leadership skill, then as it is now. In the training of U.S. military officers, many classroom hours and many more hours in Field Training Exercises (FTXs) are consumed in acquiring the skills of land navigation, determining one's location in the densest forest, knowing the most effective pathways to move troops, and knowing how to reestablish bearings via GPS, compass, map, or, as in the Middle Paleolithic, by topographic features.

The defining skill is the ability to identify (often on the move and in combat) where and how to deploy troops, how to set up a defensive perimeter, and, in this regard, how to utilize a rise of land, a fold in the terrain, a grouping of trees or other natural asset to reinforce the defensive strategy, and gain a greater chance of surviving an assault. In addition, *Homo sapiens'* innovation and advancement ultimately included superior stone-shaping techniques; needle and thread for the tailoring of highly functional layered clothing, boots, and head gear; small-game hunting and trapping; art, sculpture, musical instrumentation; and mystical or religious symbology.

From the characterizations of McGilchrist, we immediately recognize the strong left-hemispheric expression of *Homo neanderthalensis* versus the strong

Table 2.4. Hemispheres of the Human Brain: Universal Attributes

EMMISARY	MASTER
LEFT Hemisphere	**RIGHT Hemisphere**
World View	**World View**
REDUCTIONIST	HOLISTIC
Immediate objects and needs; food, sex, weapons, shelter	*The context of needs over time… strategic thinking: future-directed*
Initial Stance	**Initial Stance**
AGGRESSIVE, INFLEXIBLE	DELIBERATIVE, FLEXIBLE
Characteristics	**Characteristics**
DEFENSE OF THE STATUS QUO	INNOVATIVE AND ADAPTIVE TO CHANGE
Inability to read emotion of others	*Ability to read the emotions of others*
Impulsive Projection of Power & Control	*Strategic Planning/ (Offensive and Defensive Action)*
Rote/Repetitive Methods and Learning	*Exploratory Methods & Adaptive Learning*
Skills	**Skills**
BASIC PRAGMATIC / SURVIVAL	ARTISTIC / VISUAL
Analytical/Object Oriented	*Creative/Experimental*
Basic Language/Survival	*Context-driven/Complex language.*

Adapted from Iain McGilchrist, The Master and his Emissary: The Divided Brain and the Making of the Western World, *2009*

right-hemispheric expression of *Homo sapiens*. Right-hemisphere emergence was the tipping point toward full self-awareness, which was achieved prior to the migration of *Homo sapiens* from Africa and will be central to our further considerations. The stark differences between the life experiences and survival pathways of men and women that we have reviewed in this chapter are the baseline of our different worldviews. The fact that the right and left hemispheres embody those separate worldviews will be fully developed in Chapter Four, "The Gendered Brain," but first, we need to consider the strategic advantages of the *Homo sapiens* mindset.

We begin with the right hemisphere's superior visualization and artistic skills, which constitute a powerful competitive advantage. Nowhere in the study of art history have I found a full recognition of the military advantage that visualization skills create in describing an offensive raid, planning defensive fortifications, and communicating related combat maneuvers. However, such objectives are advanced by the high degree of specificity that can be quickly communicated by something as simple as using a stick in the sand to mark out, with clarity, the future threats and opportunities that will be encountered by a group of raiders or defenders. This objective is "A" in the alphabet of the U. S. Army's courses in Military Science.

To this day, the communication of military and athletic strategies consist of spare sketches of obstruction, topography, and movement pathways in advancing on the objective, be it the goal line, the hoop, the hilltop, or Osama Bin Laden's compound.

We witness a highly developed level of observation and related artistic skill in *Homo sapiens'* carving of the Lion Man of the Hohlenstein-Stadel cave (40,000 YBP), see Figure 4.1, and in the cave paintings of Chauvet (32,000 YBP) and Lascaux (17,000 YBP), see Figure 2.4. These high-level artistic accomplishments were not expressed, even in rudimentary form, by *Homo neanderthalensis*. Anyone who has attempted to draw with charcoal can immediately see the economy of line and powers of observation evident in the cave lions and horses of Chauvet (Figure 2.4, B and C). There is an inherent value in this ability to communicate to others the detailed nature of prey, predator, or enemy.

For the first time in history, we see a permanent and unique visual narrative of life being passed from generation to generation—an evocative common heritage, a touchstone, serving to reinforce the identity of us versus them. We tend to be overawed by the thousands of years of artwork preserved in the Egyptian tombs, but these cave paintings of Chavet Cave were created more than 25,000 years earlier! Buried deep within the Earth in total darkness, these dramatic chambers with their massive overhead images, alive in the fluctuating light of torches, would have been the essence of a moving theatrical experience, the suspension of disbelief, and would have left a collectively bonding and unforgettable impression on the observers (Figure 2.5).

A final and elegant demonstration of the level of artistic-cultural advancement of *Homo sapiens* is seen in the exquisitely fashioned musical instruments recently recovered in Germany. The creation of these instruments dates back to the time of the "replacement" of *Homo neanderthalensis* from the center of their homeland in Europe (Figure 2.6). The flute found at Hohle Fels cave, 13 inches in length, is carved from the hollow wing bone of a griffon vulture and is dated to 40,000 YBP. The flute found at Geissenklösterle cave is an assembled and adhered two-part carving of mammoth bone and is 7.4 inches in length and is dated to 35,000 YBP.

A
Detail of *Aurochs*
Lascaux Cave, France
17,000 YBP

B
Detail of *Horses*
Chauvet Cave, France
30-32,000 YBP

C
Detail of *Cave Lions*
Chauvet Cave, France
30-32,000 YBP

D
Detail of *Bison*
Lascaux Cave, France
17,000 YBP

Figure 2.4. Cave Paintings: Paleolithic Homo sapiens Cultural Expression

Collectively, the genetic and cultural findings we have reviewed in this chapter substantiate that, despite recent claims of near-equivalency between *Homo neanderthalensis* and *Homo sapiens*, there was an inherent cognitive advantage possessed by *Homo sapiens* that trumped all the evolutionary adaptations and home-field advantages of *Homo neanderthalensis*.

EMERGENCE

There is no single event in prehistory that represents the moment that humankind uniformly stepped out of an animalistic protohuman existence and into a human state of existence—a state of being fully self-aware. The process undoubtedly was distributed, uneven and occurring over long time spans. Having said that, there is no event in human history that comes as close to embodying the final test and emergence of a fully self-aware species as does the collision of *Homo neanderthalensis* and *Homo sapiens*, the two most advanced members of their genus, 42,000–40,000 YBP in Eurasia.

Whether we view this as biological happenstance or as the tragic final scene between two fated antagonists, this is the confrontation that opened the road to who we are and who we are becoming.

Genetics and archaeology, through their very different methods, are increasingly in concurrence that *Homo sapiens* had early encounters and, eventually, a surrounding presence in the heart of *Homo neanderthalensis* territory.[32] Their advance appears to have been resisted, particularly in the citadel of southern France, for hundreds if not thousands of years. This final confrontation of the "top of genus" marked the preeminence of the self-aware mind and the beginning of the next phase: human domination of the natural world.

A barely comprehensible implication here is that we *Homo sapiens*, to a high degree of probability, went on to destroy all other emerging, sentient prehuman life within the genus *Homo*. Without this bridge back to our great-ape ancestors, subsequent generations of humans lost all knowledge of the connection. Following the extinction of our last known relative from the genus *Homo* (*Homo floresiensis*), on the island of Flores, in Indonesia, it took an additional 12,000 years before a brilliant member of *Homo sapiens* (Charles Darwin) was able to reconstruct our own life story.

The question in antiquity has been:
"Where is the missing link?"

The question in the Middle Paleolithic should be:
"Why are all the links missing?"

ORIGINAL SIN

Since it is apparently irresistible to project a moral framework on the birth of humankind, what in the Paleolithic would constitute original sin? Logically it would be the assault to the point of extinction of all remaining members of the genus *Homo*. While we still carry a genetic pittance of those whom our ancestors destroyed, what sin could possibly rise to the level of such a global extinction?

Of course, this is yet another moralistic fallacy, since the *Homo sapiens* of the Middle Paleolithic were just the start of what we are still becoming. They were emerging from an internal, biological phase of animal existence and beginning an external, cultural phase in which they would eventually decide what constitutes humanity, right and wrong. There had yet to be a Moses, a Buddha, a Socrates, a Jesus, or a Muhammad. No one had drawn up the Hammurabi Code or the Magna Carta or the Bill of Rights. There were no maps, no agreed-upon national borders, no World Court, no United Nations.

If, on the other hand, we were to point to an act of redemption that marks our deliverance from such an original sin, then surely it would be the defeat of Hitler's Germany and his plans to create a Master Race. Facing this terrifying prospect for humanity, a majority of international military forces rose to defeat

Figure 2.5. Paleolithic Cave Paintings: Lascaux Hall of Bulls (17,000 YBP)

Copyright 2015 Sisse Brimberg/ National Geographic.

this "racial cleansing." While the response to the more recent genocides of Rwanda and Darfur may have been less complete, the default setting for humanity is a global condemnation of these horrific regressions. In this sense, our redemption, should we need one, is well under way.

What our ancestors possessed as a result of their travails, their mass starvations, and their brutal competition and forging was an infinitesimally small but miraculous cognitive gift. First possessed by a handful of our ancestors in Africa 100,000 YBP, this gift has now been refined and replicated across the face of the Earth by the billions. This was the starting point, the kindling of self-awareness, and through neural pathways and amplified by feedback loops, a new gateway in the evolution of life on Earth was opened. While a consensus interpretation is currently under debate, the conclusions drawn for our purposes here are straightforward:

THE BASELINE CONCLUSIONS: CHAPTERS ONE & TWO

- *Homo neanderthalensis* was forcibly replaced by *Homo sapiens* in a combination of deadly raids, violent competition, and exclusion from resources by 40,000 YBP.

- *Homo neanderthalensis* as well as *Homo floresiensis*, *Homo denisovan*, and as-yet-undiscovered prehumans and/or members of the genus *Homo* were, to the highest levels of probability, also killed outright or driven into isolation and extinction by *Homo sapiens* in the process of their "takeover" of all five habitable continents on Earth.

  The ancestral male role as the 'tip of spear' in deadly offensive attack and defensive protection of the group characterized and reinforced the worldview of the male-lineage.

- The subsequent Mesolithic and Neolithic eras (confirmed through widespread archaeological evidence) display a continuing pattern of male-directed raiding and violence among groups, characterized by the kidnapping and rape of females of childbearing age.

  This means of genetic projection and accelerated reproduction, in conjunction with female intergroup migration to avoid incest, characterized and reinforced the worldview of the female lineage.

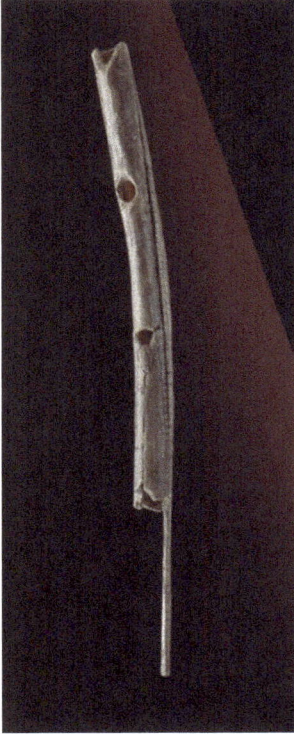 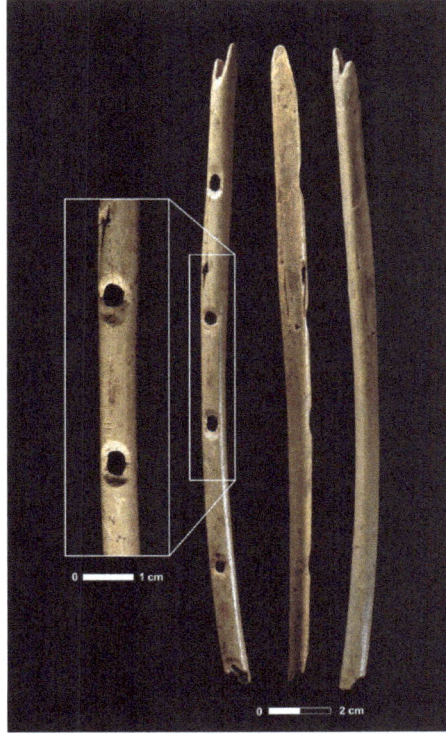

A
Mammoth Ivory Flute *(above)*
7.4 in. length
Geissenklosterle Cave, Germany
35,000 YBP

B
Griffen Vulture Wing Bone Flute
(upper right and right)
0.3 in. diameter, 13.0 in. length
Hohle Fels Cave, Germany
40,000 YBP

Figure 2.6: *Homo sapiens* Artisanship, Music, and Cultural Values

- The inherited male-lineage life experience of aggression and inflexibility characterizes the left hemisphere of our brains, and the inherited female-lineage life experience of empathy and flexibility characterizes the right hemisphere of our brains.

- The innovation of superior weaponry (logically, projectile weapons) was the cultural impetus and anchor for the natural selection feedback loop that led to the self-aware mind in all *Homo sapiens*.

    Emergence of self-awareness in the mind of *Homo sapiens* 100,000 years ago was coincident with the greater neurological integration and balance between these left and right hemispheric expressions which, in turn, was the determinant advantage in their replacement of all other members of the genus *Homo*.

CHAPTER THREE

The Necessity of God

A CENTRAL QUESTION REMAINS TO BE ANSWERED IF WE ARE TO understand the transition of *Homo sapiens* to the complexity and scale of the modern world. As noted in Chapter Two, we have physical and genetic confirmation of *Homo neanderthalensis* still living at the scale of small, isolated groups, or bands, 200,000 years after achieving dominance over Eurasia—still living within the limits of the biological bonding of the immediate family group. Why did we break out of this limitation to form large scale communities?

In one of my most memorable readings in the field of the natural sciences, I was taken aback by E. O. Wilson's compelling description of how an ant colony can function as a superorganism (an organism consisting of many organisms). Most unforgettable were the soldier ants arrayed in coordinated defense, fighting to the death, giving up their lives without a second's hesitation to protect their community and queen. Learning how their underlying genetic and chemical drivers act in concert to produce this teeming entity, this single fierce will composed of tiny life forms, I could no longer think of them as that minor nuisance at childhood picnics. Immediately, it struck me that if such life forms were to understand their own mortality and not be related to every creature in the nest, then the very first job for the continuance of that species would be to find a way to get their soldiers (possessing such knowledge) to be willing to die for the biologically unrelated thousands in the group.

If one were to break this instinctual iron-clad bond of genetic connectivity and chemical control and introduce anything approaching free will in ants, it would be every ant for itself or, as in the case of genus *Homo*, it would mean a collapse all the way back to the smallest unit of a life-or-death commitment: the

67

family group, bound by blood and sex. In such an instance, it is "us against the world," us being the band, or primitive unit typical of the *Homo neanderthalensis.*

In the Middle Paleolithic, all members of the genus *Homo* operated at this lowest scale of complexity. It was the only viable unit of security that assured the fight-to-the-death commitment that was necessary given that competing bands were frequently on patrol, looking for an opportunity to attack.

There would have to be a large scale of bonding to achieve the required life-or-death commitment beyond the scale of the family—a shared value of the highest order to be defended to the death. My assertion is that the emergence of self-awareness—the recognition of our existence within the larger universe—and the biological imperative for an explanatory narrative to address the resulting fear of the unknown, created a "necessity" for God.

In the overleaf of *God Is Not One: The Eight Rival Religions that Run the World,* Stephen Prothero reports,

> At the dawn of the twenty-first century, dizzying scientific and technological advancements, interconnected global economies, and even the so-called New Atheists have done nothing to change one thing: our world remains furiously religious. For good and for evil, religion is the single greatest influence in the world.[1]

The anthropologist Lionel Tiger and the neuroscientist Michael McGuire similarly observe in their book *God's Brain,*

> Virtually all human groups generate and sustain religions. It is simply good naturalistic science to begin and not end with that huge reality.[2]

The prehuman mind responds to the world as it is, while the self-aware mind questions why there is a world, who we are, and why we are here. In *Homo sapiens,* the self-aware mind sought and found answers to these urgent questions, which were a corollary, a necessity of this new state of self-awareness. The answers were centered on an all-powerful Creator or Animal Spirit and/or God or gods who were the central actors in their various Creation narratives. Time after time and in totally remote human groups such initial explanations were interpreted and reinterpreted into a detailed common belief system that resolved a fearsome universe and our meaningful role in it.

This proved to be a transformative concept that eventually bound together social/cultural units at the mega scale of civilizations. Without commenting on the validity of any specific religion or belief, this chapter addresses how our biological baseline demanded explanatory narratives, which, in turn, created fundamental changes in the complexity of language and the nature of leadership within human groups. Most importantly, these early, shared religious beliefs created competitively advantaged groups.

ORIGINS OF THE NARRATIVE

Before self-awareness, members of the genus *Homo* had evolved a highly successful set of predispositions for survival. Number one among these instincts, honed to a razor's edge, was the ability to quickly recognize and assess danger. This ability interprets, almost instantaneously and often subconsciously, the significance of any change in the environment: any sudden movement, sound, or vibration. Pattern recognition is a subset of this larger and multilayered sensibility.

There can be little doubt about why we tend to see agency in even the smallest changes in our environment: our ancestors who were most insightful, who could quickly assess subtle signs of danger as threat, or no threat, were the ones who lived to reproduce. The process of effectively seeing agency requires that we generate an explanation, an interpretation of the threat. This need to immediately remove uncertainty and risk operated at a biological level and was formed by the leading edge of early death.

This ability recognizes (at a safe distance) the profile, facial features, or gait of someone approaching—friend or enemy, good or evil? Natural selection quickly rewards this attribute, which is to say that those who do not have it are quickly eliminated. This is a core aspect of "fitness" because even though one may have the largest claws, the greatest speed, or the best weapons, nonetheless, in the absence of hypersensitive discernment and high-speed response, all other attributes can be defeated. Hesitation to consider ambiguity in the moment of a threat can be fatal.

A modern equivalent is found in the Basic Combat Training of the U. S. Army Infantry School. A fundamental lesson for young troops—usually communicated by combat veterans rotated out of the front lines of Iraq or Afghanistan—is what to do on patrol when suddenly coming under enemy fire. The age-old lesson is "DO SOMETHING!" That something may be to leap for cover, immediately return fire, or run in a zigzag pattern, but "DO SOMETHING!" Under no circumstances stop and try to figure out what to do next. YOU MUST REACT! In other words, *there can be no ambiguity or indecision in the moment of a life-or-death confrontation.*

We now know that there is an anticipatory function in the brain that is constantly working to inform us by running scenarios or interpretations of the signals coming in from the exterior environment, creating our moment-to-moment conscious and subconscious sense of the world and the risk it presents.[3] Many a pedestrian has been saved by the immediacy of this sensibility: reacting without consciously thinking and, as a result, not being blind-sided by a speeding car. Such interpretations need not be lengthy or detailed to serve as brief "explanatory narratives" that remove ambiguity and allow the mind to relax to a normal level of alertness or, if required, to escalate to a clear defensive or offensive posture.

Without this clarity, the brain is in a most stressful state of confusion because preparation for what happens next cannot be set. Our brain's left hemisphere is the center of this anxiety and creates pressure for immediate resolution—black or white? Recent advances in neuroscience and imaging confirm that the brain literally lights up with activity when placed in such an uncertain state. What happens when our urgent need for explicit answers, generated by the left-hemisphere mind, confronts the wide-open considerations of the other half of the brain?

The right hemisphere contemplates not only the immediate outside world within which it must function but also the inner landscape of *Homo sapiens* within which it resides as well as nature's universal trajectory of birth, life, and death within which we all reside. This level of self-awareness brings with it the contemplation of our source of being and our larger purpose within the cosmos. The brain's right hemisphere has absolutely no problem with ambiguity and the shades of gray that emerge from such considerations.

Combining the analytical and narrowly focused left hemisphere, which demands immediate and concrete explanations, with the right hemisphere, which leisurely contemplates the greatest unknowns of our existence, seems like a formula for endless indecision. However, the resolution of the two opposing viewpoints, not necessary in our prehuman internal-biological phase, was projected into an external-cultural expression. The explanatory narratives of God, gods, and religion removed the anxiety surrounding the unknowns of existence and thereby created a strong center of shared belief.

With this, there was a community of believers willing to sacrifice their lives for this larger, common purpose. Such changes had transformational effects on the nature and role of language, as well as on the structure of leadership:

THE BASELINE IMPLICATIONS

Expansion of the Explanatory Narrative: The first act of the self-aware mind is to pull back the curtain on the vast scale of the universe, on the uncertainty that exists beyond the confines of the physical, nature-based experience, to confront the realm of the supernatural. Within the individual, urged on by the left brain, such a revelation creates an immediate need to remove the new scale of uncertainty, a need to proportionally expand the explanatory narrative. In a predictable response of the self-aware mind, the biological need for the explanatory narrative is projected outward to the external realm of culture. A cultural bonding force at the scale of the metaphysical is born.

Role of Language: Within the previous nature-centered world, leadership of the band had been narrowly based on physical strength and hunting and tracking skills, as well as on leadership in migration and exploitation of new resources. Language needed only a basic vocabulary to meet these relatively simple functional needs. Now the strategic advantage of a compelling explanatory narrative became evident through the large numbers who could be drawn into the group while maintaining the willingness of individuals to die for the common good. A new criterion for leadership was opened. From this point on there would be a continuing expansion in the size of groups and an ever-widening competition for control of the explanatory narrative—a clash among competing values and cosmologies to establish leadership in the metaphysical dimension.

Language therefore emerges as a key to competitive advantage, rewarding those who possess the greater descriptive and oratorical power required to prevail in this new arena. The narrative became a life-or-death differentiator, an adaptive skill that, through natural selection, propelled language to the complex and subtle expression we now experience.

Structure of Leadership: The model of the physically dominant male leading the band of hunter-gatherers quickly became obsolete. Successful leadership of the group now had to accommodate an explanatory narrative and typically a shaman, oracle, priest, or other interface with the supernatural.

Language and its power of outreach, coupled with a compelling explanatory narrative and the physical prowess to back it up, established identity, the mega distinction between "us" and "them." Increasingly powerful and formal central belief systems worth dying for expanded, competition raged, and the prevailing human societies eventually approached the scale of a superorganism. This scalable framework (one that allows expansion to larger and larger group size) exhibited three key attributes: group coherence centered on common belief, the willingness of unrelated individuals to die in defense of the group, and the emergence of complex language.

In short, the emergence of self-awareness created an overwhelming sense of our small presence in a vast universe and exposed the limits of our understanding of complex processes such as seasonal change, celestial/ lunar/solar traverse, and, at the top of the list, our own death. This, in turn, was the wellspring of the explanatory narrative. The biological nature of the explanatory narrative—the need to impart meaning and remove uncertainty—is substantiated by the spontaneous and near-universal emergence of creation and God-centered narratives among human communities worldwide. The ferocity and longevity of such human behavior confirms that it is an inseparable feature of the self-aware mind.

MESOPOTAMIA AND THE EGYPTIANS

A dramatic exemplar of the spiritual and artistic advancement of *Homo sapiens* at the exact time of their triumph over *Homo neanderthalensis* 40,000 YBP is the figure of the Lion Man from the Hohlenstein-Stadel cave in Germany, which is carved from the ivory tusk of a mammoth elephant, as seen in Figure 4.1. The similarity to the half-animal/half-human deities of the Egyptians and Sumerians, who followed 35,000 years later, is striking.

In fact, the combining of human and animal forms as well as the deification in human form of the Sun, the Moon, and the Earth occurs in both Egyptian and Sumerian cultures. From the aforementioned Lion Man of 40,000 YBP to the Animal Spirits of Native North Americans, this melding of nature and human form appears so consistently as to have been a natural assumption of the self-aware mind.

We are now so isolated from nature that we think of it as something to be observed, such as a beautiful sunset, or as a minor inconvenience to be overcome on a cold, rainy day. However, our ancestors experienced snow, ice, rain, drought, and the life cycle of plants and animals as the ultimate forces of life and death—their life and death.

The ancient Sumerians, even before recorded history, developed a complex and intriguing construct for the universe in which they accounted for the observable world of Nature as well as the damnation of an underworld (Hell) and a beautiful and idealized upper firmament (Heaven). Their versions of Heaven and Hell were characterized as ultimate destinations where one's conduct in life would be met with just reward or punishment in the afterlife. This prescientific model of the world, so strikingly familiar, was conceived long before the Old Testament was written.

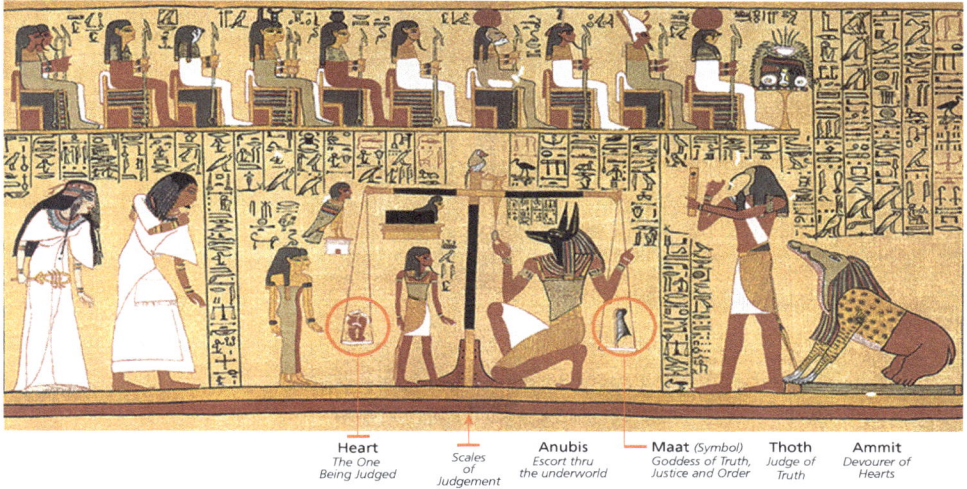

Figure 3.1. Judgment Day Before the Gods of the Tribunal

Sumeria rose to power approximately 5,000–4,500 YBP in the Fertile Crescent of the Middle East, the center of civilization's first large-scale green revolution stretching from Egypt through Mesopotamia. The current nations of Jordan, Israel, Lebanon, Syria, and Iraq fall along this arc from the Nile to the Persian Gulf: the Levant.

The starting points for recorded history are generally recognized to be cities within Sumeria and Egypt that were active trading partners at that time. Of these two great civilizations, Egypt has left the most detailed description of its religion and moral code, which also best exemplifies the adaptive value of such a belief and leadership system at the scale of the city-state. A literate and visually compelling document that captures the religious and cultural landscape of the Early New Kingdom (3,550 YBP) is *The Egyptian Book of the Dead, or Papyrus of Ani.*[5]

Contained therein are the *Declarations of Innocence*, which the deceased must proclaim before the god Ani of the Underworld during the process of the weighing of his or her heart (the literal organ) against a feather, the symbol of Maat, goddess of Truth, Order, and Justice, as depicted in Figure 3.1. A heart without sin is light as a feather. The weighing and declaration process appears quite calm until we realize that Ammit, the voracious mutant animal on the right, is poised to eat the heavy heart, sending sinners off to Egyptian Hell! Heaven was a blissful eternal life, while Hell was a bleak ending in darkness and anonymity. Personal virtue in the conduct of life on Earth was particularly challenging, since Egyptians were judged against not ten but the equivalent of forty-one commandments! The following sixteen are adequate to capture the expected standard of behavior in ancient Egyptian culture. The individual declares "before the gods of the Tribunal" that (in addition to 25 other standards) they have not sinned in the categories listed in Table 3.1.

Table 3.1. Sixteen Declarations of Innocence from the Book of the Dead

not done wrong **1.**	**9.**	not gossiped
not robbed or stolen **2.**	**10.**	not brought a lawsuit
not killed anyone **3.**	**11.**	not had intercourse with a married woman
not cheated **4.**	**12.**	not been hot tempered
not lied **5.**	**13.**	not cursed
not been sullen **6.**	**14.**	not been violent
not had intercourse with a prostitute **7.**	**15.**	not been impatient
not disputed with the King **8.**	**16.**	not been loud

My word, here is everything except Grandmother's admonition not to talk when I had food in my mouth! The origin of this elaborate model of the afterlife dates back to at least the Age of the Pyramids in the 5th Dynasty (4,400 YBP), when it was reserved exclusively for the Pharaohs. However, by 4,200 YBP it was more widely used in the form of Coffin Texts, which were available to those who could afford a sarcophagus.

The Book of the Dead, from which we draw here, dates to six hundred fifty years later (3,550 YBP), by which time it was widely understood and utilized. This date is significant when we consider that the birth of Moses was a mere twenty-four years later (3,526 YBP), according to the Zondervan Study Bible, King James Version.[6] The Exodus was dated to 3,446 YBP by Zondervan's chronology, and so, even given the uncertainties in dating these events, the Ten Commandments can be seen as an enormously abbreviated version of the professed standard of civilized conduct in ancient Egypt.

Such a standard of moral values and personal conduct is essential within the construct of the superorganism. The necessary internal consensus and collective will could never have existed had the anarchy that we now witness among the warlords of Somalia and Sudan been allowed.

While this standard creates the basis of communal harmony, the necessary militaristic commitment is captured in the following recitation before the gods, again taken from *The Book of the Dead*:

O You of the sunshine, I know the secret ways of the portals of the Field of Rushes. See, I have come, having felled my enemies to the ground and my corpse is buried.

In other words, one's obligation to the community is to obey the law, be honest, be loyal to family, and be on good behavior. The companion obligations, external to the community, are valor and sacrifice as required to defeat the nation's enemies.

The phrase "my corpse is buried" ties back to the essence of the all-embracing birth-to-death-to-afterlife intricacies laid out in *The Book of the Dead,* which provides each individual with a road map to Heaven and assures personal standing with the gods. In this manner, the state and the priests possessed a powerful form of glue that bound the nation together. After each key segment in *The Book of the Dead* are instructions, for example, "One shall utter this Chapter [text] pure and clean [bathed] and clad in white garments and sandals, painted with black eye-paint and anointed with myrrh..," followed by the phrase, "a matter a million times true."

The extensive pantheon of Egyptian gods, combined with the highly formalized and artistically rendered passage into an afterlife of Heaven or Hell, represents one of the most ingrained and tangible narratives ever achieved. The elaborately detailed procedures for the treatment of not only the bodies of the dead, but even the body's major organs in the meticulous embalming and provisioning for their journey into the afterlife, leaves little doubt as to the central role of the narrative in their culture. The highly detailed scripts that were written prior to death to assured success in interrogations by the gods created an effective intimidation of the living: preparing for the ultimate final exam!

One is hard-pressed to find a more complete example of an early human society operating so closely to the effectiveness achieved in a superorganism. While the highly developed religious/moral/ethical narrative could not, alone, create such a society in the ancient world, it is difficult to imagine how Egypt could have endured for so long, absent such a compelling explanatory narrative. In this instance, it was required of Pharaoh and commoner alike and was experienced as a seamless common vision and belief. The sheer scale of built works, stunning excellence and refinement of art and architecture, advancement in the sciences, and the military and cultural dominance of the Egyptian civilization over millennia serve witness to a powerful and multigenerational sense of cohesion and common purpose.

In Egyptian law and the conduct of society, there was also an uncommon level of equality shown to women, even in leadership of the nation, a balanced view that continues to elude many nation-states, including the United States.

Another distinctive feature was the prohibition of infanticide. Neither this level of equality for women nor the respect for each human life born into the world was achieved by Greece or Rome. These are markers of an emerging humanism.

RISING TO EDEN

As an exemplar of the explanatory narrative, the Egyptian model has a number of advantages, not the least of which is that it flourished prior to the Abrahamic religions of Judaism, Christianity, and Islam. In this way, we avoid the distraction of modern interfaith controversies and can get to the underlying universal qualities.

The most important question is how did such a detailed moral framework appear, fully formed, at the threshold of recorded time? In fact, we can be confident that it was not a new creation by the Egyptian or Mesopotamian civilizations but, rather, that these values were ancient, a fundamental biological sensibility that was progressively built upon.

Such moral values, or rules of personal conduct, may have been exploited on occasion by those seeking personal power or influence, but to be effective, they would need to have deep resonance with a strategic segment of the power structure and the community so governed, or the competitive advantage, the binding power of such a state would be eroded.

Empathy, the ability to see and feel the joy or sorrow of others, stands much farther back in time. In this phenomenon we find an inseparable partner of the self-aware mind and the foundation of a universal moral framework. The neurological mechanism for this ability is captured in the phrase "Monkey see, monkey do" and is the province of the mirror neurons of our brain. While it is the essence of our ability to learn from others, this feature operates at an even deeper level, which allows us to experience the frightening events in a horror movie or the sexual arousal felt when reading *Lady Chatterley's Lover*. We participate on an emotional level even though we are not physically participating in the event.

Despite the scarcity of the fossil record, we find clear evidence of empathy in the life of *Homo neanderthalensis*: for example, the remains of an individual with a deformity who lived to maturity and another with grievous wounds who lived far beyond the date of the injury, as evidenced by the advanced healing of damaged bones. Both situations suggest the nurturing and empathetic support of others. We see additional instances of empathetic behavior today in the wild among chimpanzees and great apes, suggesting that this was a deep ancestral quality in the genus *Homo*.

However, with the advent of self-awareness, a more intense level of empathy flows from this new perspective—standing outside the narrow immediacy of events and able to observe the suffering of fellow humans, as well as understanding the shared risk, dangers, and reversals that often ended in an early death. Eventually finalized as a consensus of personal conduct, even such an "ungodly" group as the New Atheists has recognized the existence of "a universal and objective secular moral standard."[7]

Concurrence on the universality of a moral sensibility across all nationalities and the vast majority of religions lends credence to the existence of a secular moral standard which can impart maximum binding power when woven within a given culture's explanatory narrative.

Early narratives were God-centered and surrounded by some version of spiritual formality, but religion has increasingly moved away (and sometimes has been banned) from the explanatory narratives of the most advanced and productive nations of our time. To imagine that formal religion, even in its heyday, was somehow creating wars and acting as a force of evil in the world independent of the self-serving elites who often influenced it for personal ends (a force still at play in the modern world) is simplistic. A narrative prevailed because it was effective in binding a community around consensus belief and facilitating collective action—period.

A brief examination of three national explanatory models demonstrates the range of such narratives. The human sacrifice practiced by the Mesoamerican cultures of South America and the atheistic state order of the Soviet Union are at the extremes of variance. The Vatican and Papal States of the Renaissance, on the other hand, represented the height of religious conformance. All three models, I would argue, achieved a consensus moral standard and all three were subject to manipulation from leadership for personal enrichment and power.

The Vatican and the Papacy during the Italian Renaissance constituted the center of a broad moral/ethical popular consensus, positioning religion at the center of power within the State (indeed, the church's own State). As primary author and interpreter of the explanatory narrative, the church amassed enormous wealth and power and served as patron for the artistic and scientific endeavors of others who would further reinforce that narrative.

Therefore, when Galileo endorsed the Copernican proposition of heliocentrism, that the Sun and not the Earth was the center of the universe, it was received as unwelcome news of the burned-at-the-stake variety.[8] Galileo's action constituted encroachment by independent scientific thought on vested power, a challenge to the control of the narrative, which, in this case, was seen as both religious and scientific. Galileo was ultimately coerced into recanting his statements and barely got away with his life.

The Soviet Union was the first state to eliminate religion as an ideological objective. State atheism, or *gosateizm*,[9] counteracted the perceived conspiracy of capitalism and religion as elitist partners in the exploitation of the working class. The moral and just objective of this new narrative was to achieve a society with no division by class or alienation of the individual. By having common ownership of the means of production, humankind was to be free of oppression and scarcity. This Marxist-Leninist ideal did not come to pass in its dictatorial and (post-U.S.S.R.-collapse) oligarchic reality. However, when viewed in the context of the czarist monarchy and the related suffering of the Russian people, the equality of an "all for one and one for all" explanatory narrative had an undeniable appeal.

Mesoamerican Society: The Incas represent an alternate narrative in the extreme with their practice of human sacrifice to appease the gods. Much like the parable of Abraham and Isaac, the Inca would give up one of their most cherished children to be transformed, for example, into the spirit of a mountain.

Rather than the evil act that we may see, such sacrifice was for the Inca an act of atonement to the gods on behalf of the larger community, often accompanied by celebration of the selected child and his or her adornment with elaborate garments and drugging to minimize pain. This was apparently considered a highly honorable death.[10] Recent translations support the notion that the selected or volunteering individuals considered this to be a gift to the community in the same sense that soldiers die for their country.

In the foregoing examples we see both religion and anti-religion constituting "banner" narratives, but each of them is nonetheless wrapped in a culturally grounded set of values. These are legitimate, that is to say, consensus values that resonate with the governed, even if they are abused subsequently by a small elite for personal gain. Such abuses by leadership undercut the competitive potential of the nation. Undermining the national strength of belief undermines the commitment of the individual, the soldier who is defending the nation and who must be relied upon even in the face of their ultimate sacrifice.

Since this chapter is titled The Necessity of God, it seems only right that we pause to briefly consider the arguments of the New Atheists. Collectively, the four atheists in opposition—Daniel Dennett, Richard Dawkins, Sam Harris, and

Christopher Hitchens (Hitchens died in 2011)—avoid the greater population of people who believe that there is a God by focusing on the smaller and more vocal community of literalists, those who believe that the Bible is word-for-word correct and the word of God.

While this has proved to be a great debate strategy, it ignores the majority of believers, many of whom do not regularly attend churches, synagogues, or mosques. This majority sees the Bible as a uniquely influential document composed over time by mortals who, inspired by a belief in God, wrote a series of narratives, parables, and poems compiled and edited by others over many lifetimes. Setting aside the fish-in-a-barrel criticisms based on the Bible's inherent inconsistencies and the scientific impossibility of miracles, it is instructive to consider some of the stories that most people consider parables, but which have become targets of the New Atheists.

A favorite example comes from the first book of the Bible: Genesis. From this prescientific period, darker than the Dark Ages, we have been handed down a highly improbable description of a singularity, an instantaneous act of God's creation of the Universe out of the VOID! Now, thousands of years later, we have the current scientific model of the Big Bang, a singularity, an instantaneous event that created the universe out of the VOID. As a shot in the dark by believers in that remote pre-scientific period, a surprisingly good analogy.

Next we turn to the Garden of Eden, and (if we can get past the temptation to dwell on talking snakes) we see that this parable essentially says that we began our earthly existence in a moment of recognition between right and wrong. We came to know sin, which, of course, requires the recognition of what is not sin: moral conduct. For the purposes of my argument, this is an excellent metaphor for the dawning of self-awareness, which, in turn, is inseparable from the awareness of the universe and the consistently associated explanatory narratives of religion and universal moral sensibilities. Again, not a bad analogy.

Even the creation of Adam first, followed by the creation of Eve from Adam's rib, can be stretched to have a biological parallel. The first 100,000 years of *Homo sapiens'* existence, as discussed previously, was an animalistic state dominated by males and their left hemisphere of the brain. Males and females only emerged as human, in the second 100,000 years of our species, with the development of self-awareness: a rebalancing and integration hemispheres of the brain—the foundation for deliberative thought.

Empathy, which existed far back in mammalian evolution and the genus *Homo*, was deepened by the emergence of self-awareness. It evolved, through the competition for the explanatory narrative and written language, into the much more elaborated and formalized moral code that we see embodied in most religions today. In short, religions are an extension and cultural elaboration of prehuman biological dimensions. This provenance is about as authentic as one could hope for in a human cultural expression.

A point of agreement: this author shares the New Atheists' frustration with the large segment of the population that continues to deny the reality of evolution and natural selection (which, ironically, would have come as no shock to the majority of humans who have walked the Earth). The Abrahamic religions have heightened this alienation from nature by stating that humans are made in the "image of God" and that nature is specifically demoted to a near afterthought over which humans are given dominion.

The Second Commandment of the Bible states specifically that there should be no idols of birds of the air, animals of the land, nor fish of the sea. This denial of, or more accurately, failure to acknowledge our interdependence with nature, has unfortunately stuck, remaining a signature feature of our modern culture even as religion has lost its central or even adjacent position to national power structures.

Amazingly, within the contemporary anthropological community, there is a popular and politically correct Garden of Eden theory of our hunter-gatherer beginnings. Essentially a tale of long-lost gender equality, it describes a sort of "noble savage" era of equitable humanity that was destroyed with the advent of civilization. Agriculture and the emergence of cities, with their fixed settlements and accumulation of surpluses, inciting possessiveness, thievery, and gender inequality, play the role of the villain. The discovery of defensive perimeter walls in many of these early agriculturalist settlements is pointed to as proof of the "emergence" of violence, as if our animalistic and violent transitions up to that point had never happened.

A key justification for this view is drawn from the study of selected nineteenth- and twentieth-century hunter-gatherer communities, all of which are in remote locations as would be expected in cultures still existing at near-Paleolithic levels. The relatively low violence and high gender equality of these communities are interpreted as confirmation that these inherent human qualities existed before the corrupting influence of civilization. However, as Lawrence Kelley's work confirms, to the extent that there are exceptions to violence and warfare among primitive cultures, these exceptions tend to occur among the tribes or bands that are so remote as to be isolated from potential aggressors. In fact, they may have chosen their isolation specifically as a retreat from violence, to avoid such contact.

There is an important insight to be drawn, but it is not the existence of gender equality in prehistoric times. Rather, it is that the removal of the constant threat of violence (either by retreat to peaceful, isolated settings, or by military equilibrium) creates the opportunity for the empathetic, creative, and humanistic side of our human natures (femaleness) to gain prominence. These hunter-gatherer communities, through their various forms of isolation, have achieved and maintained a nonviolent setting in which there is the breathing room for a natural upward evolution toward a more equitable society even as they remain at a subsistence level as the result of their retreat from the centers of competition. Modern civilizations achieve this secure precondition for upward evolution to more equitable societies through military equilibrium, or strategic agreement.

The evolution to increasingly nonviolent cultures has been made visible through the studies of Lawrence Kelley and others, cited in Chapter Two. Their insights were embraced and expanded in Steven Pinker's recent book *The Better Angels of Our Nature: Why Violence Has Declined*.[11] The inverse proves the point because as we go back in time, we enter an increasingly violent environment until we finally reach Africa over 100,000 years ago, our animalistic baseline.

The Garden of Eden of the Bible can be seen, metaphorically, as the starting point, the first step in the emergence of our humanity, and I would assert that the dawning of self-awareness marks such an emergence. However, rather than being thrown from the Garden, we began our rise to our increasingly nonviolent Garden of Eden. This ongoing phenomenon is deeply grounded in gender, which remains the most personally consequential and yet consciously avoided topic of inquiry in the sciences. A new conceptualization of gender and its role in forming our unique worldview is the business of Chapter 4: The Gendered Brain.

CHAPTER FOUR

The Gendered Brain

GENDER IN THE PHYSICAL SENSE SETS THE WAY THE WORLD SEES US, whereas gender at the level of the mind sets the way we see the world. The mind is the center of our sexual preferences, identity, and, unknown to most, a wide range of predilections, talents, and skills that are independent of our sexual dimensions. As important as gender is, there is no other aspect of being human that is so misunderstood. There are many reasons for this, not the least of which is the intimate and socially sensitive nature of our gender and the fact it's hard to be objective when you are born with a biological conflict of interest. For all these reasons, research in this emotionally charged area of the natural sciences is consistently underfunded and underrepresented. Indeed, the very meaning of the word is contentious.

Gender, as defined in our considerations here, is biological and is expressed in two ways. Anatomical gender is our external reproductive identity, whereas neurological gender is our internal ancestral legacy expressed as a blend of "maleness" and "femaleness" in the respective hemispheres of the brain: our "two minds." We are anatomically male or female, but in our perceptions and behavior we are male *and* female.

The unique survival pathways for males and females were handed down— through natural selection—as the DNA "sourcebooks" of male and female viability. Our human natures, our forms of societal order, and the resilience of our species are grounded in these baselines of male and female survival. As we begin to address the role of gender in the formation of our mindsets, we will also address the forms of bias and political correctness that plague the subject.

Our modern world is inundated with gender-centered complexities and misconceptions. We begin with the following short but memorable conversation:

> The building contractor, upon hearing the young man identify himself as an architect, replies "So, not gay enough to be an interior decorator but not man enough to be a contractor!"

Although dismissible as a crude joke, this comment contains a couple of backstory observations that are true. There are professions that have a disproportionately high ratio of homosexual men, as well as ones with a disproportionately high ratio of heterosexual men.[1]

Paul R. Ehrlich, professor of biological sciences at Stanford, had a number of relevant observations in his excellent book *Human Natures: Genes, Cultures, and the Human Prospect*:

> [B]ecause of environmental factors…there is in a sense a range of "genders" in human beings—not a series of sharp male-gay-lesbian-female boundaries.[2]

Reviewing a case-in-point, Ehrlich observes:

> This "clearly supports the notion that many male-female differences in sexual behavior are physiologically determined by pulses of hormones during prenatal development and are not simply under the control of hormonal states or social learning later in life."[3]

Ehrlich further offers a constructive observation about the physiological nature of resulting sexual orientations and the unlikely option of individuals' being able to "choose" to change such an inherent personal characteristic. He then wanders into societal critique:

> Understanding of this biological result would seem to have policy relevance because people seem generally more tolerant of homosexuality when they believe that it is an accident of birth.[4]

How could a repetitive pattern of variability in sexual attraction and identity ranging from 3.5 to 4 percent of the general population,[5] combined with wide-ranging individual expressions of maleness/femaleness independent of anatomical gender (also seen across the population at large), be perceived as an "accident of birth?" Surely this is the reflection of an underlying framework of evolutionary fitness. It is simply too coherent a phenomenon and too consistently present not to be functional. When I first read Ehrlich's book in 2002, I was not aware of any supporting hypothesis on the subject, but I began to consider the nature and logic of such a theory.

MATTERS OF DIFFERENCE

The core reality for both sexes is that we have both graduated from our animalistic baseline as great apes and have risen, by a miraculous expansion of consciousness, to self-awareness and a high level of self-direction. Males and females have demonstrated the capacity to excel in all roles and all aspects of civilization; this acquired mastery defines us as modern humans. Self-awareness has freed us to stand outside ourselves, to examine our physical residence: the body, bones, blood vessels, and brain that we have taken over for a lifetime. Through dispassionate self-assessment we can better understand our individual predispositions, our inherited physical and mental capabilities—male and female in origin—which will free us to live our lives with a greater sense of our talents and potential.

First we turn to an individual who exemplifies such personal mastery, one who has exhibited rigorous indifference to the cultural and societal forces of political correctness that have shaded more than a few findings in the field of human origins. Louann Brizendine, M.D., is an endowed professor of psychiatry at the University of California, San Francisco, and formerly on the faculty of the Harvard Medical School. Upon writing her first book, *The Female Brain*, in 2006, Brizendine observed:

> In the 1970's at the University of California, Berkeley, the buzzword among young women was "mandatory unisex," which meant that it was politically incorrect even to mention sex difference.

> The biological reality, however, is that there is no unisex brain…I have chosen to emphasize scientific proof over political correctness even though scientific proofs may not always be welcome.[6]

In 2010, Brizendine wrote her companion book, *The Male Brain*, which completes a great pairing of source books on the general natures of the male and female brain.[7]

The differences in structure and relative emphasis of constituents that Brizendine identified in the frontispiece of *The Female Brain* (page xiii) and *The*

Male Brain (page xv) are summarized in Table 4.1.

Reading the left side of the chart, one might say that all men make good protectors of family and defenders of territory but that they are impatient, aggressive, and overly competitive. I might also add that they are overly optimistic risk takers and will push sexual advances in the absence of a woman's interest.

Reading from the right side of the chart, one might say that all women are indecisive but are empathetic and insightful in evaluating others. One might also add that all women worry unnecessarily and hold a grudge.

The answer to whether these are absolute characterizations is an unequivocal NO! Individual men and women range across all the positive and negative interpretations that could possibly be drawn from this list and cannot be prejudged. Golda Meir and Margaret Thatcher never shied away from the strongest national assertions of territorial claim, nor did Gandhi and Schweitzer fail to sacrifice their all in behalf of the suffering of others. They reflect the diversity of our worldviews and the individual's power to shape and direct their human potential, a central tenet of this book.

The reality is that there are inborn inclinations, predispositions, and, as Iain McGilchrist has expressed so well, "ways of being in the world," most of which we may choose to embrace, reject, or refine. Insight into what those respective "inheritances" are allows us to understand our greatest strengths and weaknesses and, through self-knowledge, achieve greater levels of self-realization. The first effect of anatomical gender/DNA is to physically shift approximately half the population to baseline anatomical gender that is either male or female, but this does not remove the variability of neurological diversity that sets each person's unique perception, capabilities, and ranges in the interpretation of gender.

Brizendine, in addressing sexual orientation and the male brain notes the findings of anatomical and functional differences between gay and straight brains. Among her findings:

> A part of the hypothalamus called the suprachiasmatic nucleus (SCN) is twice as large in gay males when compared to straight males.[8]

> [Author's note: This has been proposed by some as a bisexual influence.]

> The anterior commissure—a bundle of super-fast cables that connects the brain's two hemispheres—is larger in gay males than in straight males.[9]

> (Note: This structure, which is also larger in women than in men,

Table 4.1. Inclinations and Emphasis in the
Making of the Male and Female Brain

MALE BRAIN (Larger and/or More Capable Aspects)	FEMALE BRAIN (Larger and/or More Capable Aspects)
"Defend your turf" DPN Dorsal Premammilary Nucleus (one-upmanship and territoriality)	PFC The CEO of the Brain Prefrontal Cortex (good judgement, inhibits impulse)
"Threat: Alarm System" AMYGDALA Temporal Lobe (drives emotional impulse)	MNS "Emotional Empathy System" Mirror-Neuron Systems (reads non-verbal emotional cues)
"Sexual Pursuit" MPOA Medial Preoptic Ave (2.5 times larger in male)	ACC "Worry-wart Center" Anterior Cingulate Cortex (weighs options, hesitates before decisions)
Cognitive Empathy TPJ Temporal Parietal Junction (races towards "fix-it-fast" solution)	INSULA "Processes Gut Feeling" Cerebral Cortex (reads intention of others)
Motivation Center VTA Ventral Tegmental Area (rewards movement/action)	HIPPOCAMPUS Verbal/Performance Memory Limbic System (never forgets emotional events)

is believed to be involved in sex differences related to cognitive abilities and language and fits with the finding that gay males, like females, have better verbal abilities than straight males.)

Anatomical asymmetry in the size of the two brain hemispheres that is characteristic of straight male brains is not observed in gay male brains. Instead, their magnetic resonance imaging (MRI) studies showed that in this respect (brain morphology) gay male brains were more like female brains.[10]

[PET scans confirm that] [T]he connectivity of the amygdala of the gay male brain is more like that of the straight female brain than of the straight male brain.

Straight men outperform straight women on tasks requiring navigation…[G]ay men perform more like straight women on such tasks.[11]

A notable aspect of these findings is not just the wide-ranging physical and functional differences that characterize the brains of gay men but that, as

Brizendine observes, *other significant functional and anatomical differences exist that are consistently present but "not directly involved in sexual attraction."* The earlier observation, that there is a wide range of maleness and femaleness across the population independent of anatomical gender, suggests that homosexuality is just a degree of gender expression within a larger pattern of neurological gender variance. Most important, there appear to be other nonsexual dimensions of diversity; talents, and skills that are closely associated with these pathways of gender variance.

The neurological gender programming process that occurs during the prenatal/ neonatal development of brain cells has been well described in an endocrinology study by Dick F. Swaab and Alicia Garcia-Falgueras, published in 2012,[12] relevant assertions include:

- Fetal sex hormones and genes interacting with developing brain cells sexually differentiate the brain and establish gender identity (male or female) and sexual orientation (heterosexuality, homosexuality or bisexuality), which are largely determined for life.

- The sexual organs (anatomical gender) are differentiated in the first two months of fetal development, whereas the brain (neurological gender) is differentiated later, after four months.

  [Author's note: Therefore, the neurological gender may not be in alignment with the anatomical gender or may even be counter gender, as in the case of homosexuality and transsexuality.]

- Sexual orientation does not appear to be caused by or reversible by social or environmental factors.

A characteristic variance between the brains of men and women, as well as heterosexuals compared to homosexuals, is in the "lateralization" of the two hemispheres, that is to say, the nature, functionality, and frequency of connections between the hemispheres. Those connections appear to be primarily contained in the corpus callosum, a communication highway between our "two brains."

COLORATION OF THE BIHEMISPHERIC BRAIN

Iain McGilchrist, in the introduction to his extraordinary accomplishment of science and scholarship, *The Master and His Emissary*, notes the link between cerebral lateralization and creativity and goes on to say that it "may also be associated with homosexuality, which is thought to involve a higher than usual incidence of abnormal [atypical] lateralization."[13] Similarly, he observes that a number of mental illness genes and patterns of lateralization associated with bipolar disorder and autism are also associated with creativity and savant gifts and would "have been bred out long ago if it were not for some hugely important benefit that they must convey" in the population as a whole.[14]

The observations of McGilchrist and Brizendine on the nature of lateralization imparted in the developing fetal brain and the range of nonsexual capabilities that are selectively swept up in a given neurological gender seem to point to a more comprehensive organizational effect.

McGilchrist explores in captivating detail the nature of our bihemispheric brain and the two fundamentally different versions of the outside world that are delivered to us. The following observations, more explicit than those in Chapter 2, serve to elaborate on his theory:

- Each hemisphere of the brain is interpreting what is going on in the external world differently and these two versions are actually different "ways of being in the world," creating a power struggle between the hemispheres of the brain.

- The left hemisphere, which possesses an "excessive and misplaced rationalism," sees the world as pieces of information in isolation and is overly analytical and reductionist, whereas the right hemisphere tends to see things in context, as a whole, and possesses a keen sensitivity to the new, creative, and innovative.

- Aggression and anger are emotional inclinations of the left hemisphere, whereas empathy, depression, and trust characterize the right hemisphere.

There is no way to capture here the breadth of considerations and insight achieved within the nearly five hundred pages of *The Master and His Emissary*, but in Table 4.2, I have created a comparative chart to summarize various characterizations used by McGilchrist in describing the nature of each hemisphere and have grouped them in categories for our purposes here.

Interpretation

Going beyond the characterization of trust and no trust observed in Chapter 2, we now recognize a more developed portrayal of each hemisphere:

Left Hemisphere	=	Protect and Defend (All Doubt, No Trust): SURVIVAL The Goal is to protect the Status Quo
Right Hemisphere	=	Expand and Cooperate (All Trust, No Doubt): OPPORTUNITY The Goal is to grow beyond the Status Quo

Assertion

In the Introduction to *The Master and His Emissary*, McGilchrist notes:

> Perhaps the most absurd of these popular misconceptions is that the left hemisphere, hard-nosed and logical, is somehow male, and the right hemisphere, dreamy and sensitive, is somehow female.

Far from a simple male/female designation for each of the hemispheres, the model that I am proposing is a multiplier of gender complexity, which stands in direct contrast to the concept of the universal model that McGilchrist further describes as applying to 95 percent of the population.

The basic premise of my argument is that there is no typical cerebral organization and certainly not one with 95 percent currency, given the role of RNA. Rather, there is created, across the population, an advantageously diverse range of cerebral organizations and related lateralization at a level of complexity that we have yet to fully appreciate.

As mentioned earlier, the human genome has approximately 22,000 genes, with only a couple of hundred genes related to the development of the neurological system, brain, and therefore, behavior. The premise of *Two Minds* is that there are two complementary forces at play in this critical area.

1. *Protein coding DNA*, which, over thousands of years, acting through natural selection, passes down the "maleness" and "femaleness" hemispheres of the brain—designated here as DNA.

2. *Non-protein coding RNA (ncRNA)*, which, during the process of birth, imparts a unique interconnectivity between the hemispheres of the brain—designated here as RNA.

RNA can therefore be seen as operating at the level of the formative mind in a process of *neurological gender synthesis* (NGS). The complexity potentials, the range of human personalities that arise from this role of RNA, are nearly infinite.

Table 4.2. Characterization of the Hemispheres of the Brain

"Emissary" **Hemisphere L**	"Master" **R Hemisphere**
WORLD VIEW	**WORLD VIEW**
OBJECTS, NUMBERS, WORDS (Reductionist)	Holistic, the WORLD in CONTEXT (Inclusiveness)
INITIAL STANCE	**INITIAL STANCE**
Frontal lobe BETRAYAL EXPLOITATION	Frontal lobe TRUST EMPATHY
CHARACTERISTICS	**CHARACTERISTICS**
AGGRESSION ANGRY TERRITORIALITY COMPETITION, RIVARY MECHANISTIC INFLEXIBILITY to read emotion of others INABILITY EXPLOITATION UNIFORMITY See another point of view UNWILLING UNDUE Cheerfulness STATUS QUO INTOLERANCE PROJECTION OF POWER	HUMOR SOCIAL ALTRUISM COOPERATION HUMANESS FLEXIBILITY ABILITY to read the emotions of others BONDING (Genuine) UNIQUENESS WILLING to see another's point of view UNDUE Sadness/Depression INNOVATION SENSE OF JUSTICE/Righteous Indignation SPIRITUALITY
FEARS	**FEARS**
ANXIOUS APPREHENSION Fear of uncertainty Fear of lack of control	ANXIOUS ARROUSAL Reading a sad story = sorrow Melancholy–sadness response
SKILLS	**SKILLS**
TECHNOLOGY Performance/MUSIC Linear MATHEMATICAL REASONING Execution/Manipulation of NUMBERS LANGUAGE/VOCABULARY	SPACIAL Interpretation (3-D) MUSIC/Innovation and creation ARTISTIC/VISUAL COMPOSITIONAL/Insights CREATIVE WRITING and VERBAL INTONATION
INTERESTS	**INTERESTS**
ACQUISITION OF MATERIAL MEANS ORGANIZATION/BUREAUCRACY PLANS, MAPS, STRATEGIES INFORMATION and Gathering	ATTENTION to: SEX, ATTRACTIVENESS and AGE FOOD/SOCIAL INTERACTION LITERATURE MUSIC and the Arts

Adapted from Iain McGilchrist, The Master and His Emissary: The Divided Brain and the Making of the Western World

I have designated the result of this synthesis as a unique *neurological gender ratio* (NGR)[15], an individual's unique pattern of sharing between the two gender-informed hemispheres of the brain.

The beauty of this model (male legacy/female legacy) is that it goes back to the core logic that different selection pressures on males and females are represented in the two halves of the brain as the collective knowledge of humankind. Nature thus conserves and projects both forms of fitness into the next generation. The realization that our personal subset of these two legacies exists in the physical form of the hemispheres of our brain and that there is an internal conversation that we experience as a three-way debate between those male and female ancestral predispositions and our feedback from the world we are living in—our free will—is profound. Those people who believe in reincarnation or feel that they have memories of past lives may have a greater basis for their claims than we ever knew!

I propose that there is no "taking over" of the left or right hemisphere influence in any culture but, rather, that an initial range of relative influences between the hemispheres, NGRs, are imparted across a given population and that the proportions change over time in response to the unique environmental conditions. For instance, the proportions of left or right hemisphere influence can subsequently rise or fall in just a few generations in response to the challenges of survival (left hemisphere rise) or the rewards of opportunity (right hemisphere rise).

In reading the personality of the right hemisphere, there is an elemental theme of empathy and emotional intelligence; therefore, for our purposes, the general suite of characteristics is designated as *femaleness*. (Only one trait located in the right hemisphere—spatial/navigation—is expressed in males, but this exception serves to prove the rule and, in any event, does not change the overall character of the hemisphere).[16] By contrast, in reading the personality of the left hemisphere, there is an elemental theme of aggression, territoriality, and lack of emotional intelligence; therefore, for the purposes of our assertion, its suite of characteristics is designated as *maleness*.

The specific assertion is that in the prenatal and neonatal process, through RNA interpretation, there is a maleness/femaleness ratio (a uniquely wired or selectively inhibited pattern of functional relationships between the two hemispheres of the brain). For instance, in the male range, an alpha male would be at one extreme and an omega male—transsexual—at the other.

Consistent with Swaab and Garcia-Falgueras's study, an anatomical gender assignment (sexual organs) occurs first, and then, later in the process of development, a unique neurological gender interpretation is created as an overlay in the fetal brain. Each interpretation exists as a subtle functional and sometimes morphological variation within the hemispheres of the brain.

We all have friends and family who come to mind when we observe this inherent human variability, and we need look no further than presidents of the United States and candidates for that office to see the same variations.

It is not going out on a limb to say that Jimmy Carter and Al Gore, in their compassionate and empathetic qualities, are more femaleness-expressive than George W. Bush and Ronald Reagan, whose higher maleness-expressive concerns center on the military and our national defense.

John F. Kennedy and Bill Clinton, on the other hand, are approximations of 50/50 NGR, an equal balance of maleness/femaleness and, not surprising, with their ability to relate to many different people, they are the most popular United States presidents of the last half of the twentieth century.

While such a wide-ranging and subtle biological variance in the brain has yet to be detected, its presence can be inferred from the more visible variation in lateralization associated with gender, savant, autistic, and psychological anomalies remarked on by Brizendine, McGilchrist, and others. The existence of such an underlying neurological diversity is consistent with the intractability of such deficits as lacking a sense of humor, a sense of rhythm, and emotional intelligence or sensitivity. It places sexual orientation and preference (heterosexual, homosexual, and bisexual) as well as sexual identity (transsexual) within a coherent model or neurological sexual spectrum.

A related outcome is the replacement of the current assumption that the proper functioning of the prenatal/neonatal period of brain development, via genes, hormones and other vectors, should create a neurological gender that matches anatomical gender and that all other outcomes (most visibly the cross-gender outcomes) are misfires, abnormalities, or disorders. My fundamental premise is that the existing prenatal/neonatal process and outcomes create a high level of evolutionary fitness and that we currently fail to recognize the diversity and species survivability being imparted.

DOWN THE RABBIT HOLE

Before we attempt to map out biological pathways and the phenomenon of neurological gender synthesis (NGS), we need to address a couple of gender-based distortions of long standing in the areas of the social sciences.

It has not gone unnoticed, I am sure, that all references to the emergence of the right brain, or "femaleness" dimension, and its powerful significance in the emergence of *Homo sapiens* up to this point has been through the actions and accomplishments of males (even in the comparative example of U. S. presidents and candidates for that office). While there have been strong women leaders with unquestionable "maleness" attributes—Margaret Thatcher, Golda Meir, Amelia Earhart, and Eleanor Roosevelt come to mind—the reality is that from the beginning of our animalistic baseline, there has been a glacial thawing of the near-monolithic domination by males in human society. Only recently has society been moving to full equality for women, gays, bisexuals, transgenders, and the wide array of racial minorities across our population. Same-sex marriage has only recently become legal in all states of the United States (2015).

Women are now leaders of many nations around the world, and Hillary Clinton or Carly Fiorina may—and some woman will—become president of the United States in the near future. The current leadership of women CEOs, senators, governors, news anchors, scientists, commentators, and editors all confirm that a long-overdue rebalancing of the relative influence and standing of women is well under way in the western world. This rise in standing applies to both anatomical females and anatomical males with femaleness-dominant NGRs, which is consistent with the 100,000-year trending we are following.

Nonetheless, a host of politically correct distortions exist that bear directly on the assertions under discussion. Nowhere is this more evident than in the whitewashing of the overall male/female dynamic of the Middle and Upper Paleolithic, a favorite arena for such distortions in the social sciences.

When researching the scientific literature for this emergent period of 20,000–30,000 years before recorded history, one encounters twentieth-century concepts and characterizations that are simply not relevant. First among these is gender equality. A visit to just over 200 years ago should make the point. The primary author of our Declaration of Independence was a wealthy white male slave owner from Virginia who, at the behest of his fellow all-white male representatives of the original thirteen states, drew on concepts of the European Enlightenment to make the case for our independence from England. He penned the words: "All men are created equal."

Starting from that imperfect and aspirational statement, our interpretation of those words has evolved upwardly over the last 239 years to include the abolition of slavery and the recognition of equal rights for women, people of color, and, most recently (and ongoing), the equal rights of gay, bisexual, and transgender members of our society. Eventually, the words may appropriately read: "All people are created equal." In the context of a civil society, we have defined, legislated, and only recently begun to make real the full equality of rights, an equality that we regularly interpret, adjudicate, and enforce.

How can we apply such concepts to the Middle and Upper Paleolithic context of 40,000–22,000 YBP when we find it difficult to project them back just over 200 years? With apologies for a simplistic overstatement, we don't say that a lion murders a gazelle; we understand that there is an evolved animalistic state in play. At the first dawning of full self-awareness, engulfed in an animalistic struggle for survival, our ancestors, male and female, drew on the inheritance from their prehuman ancestors. Their conduct was the sum of a brutally tested mode of survival that worked. They were successful to the extent that, just barely, we are here today.

The path to modern humanity is not something for us to judge by today's standards; neither is it something that we are doomed to repeat, nor is it a basis for guilt. It was a successful path on the way to becoming human, and that's enough.

Just as we have crowned the African lion the "King of the Jungle," there can be little doubt as to why our Middle Paleolithic ancestors chose to embody the European cave lion in human form. Unlike the mammoth elephant, cave bear, bison, and aurochs, which were larger and arguably stronger, the cave lion was a human-sized, cunning, and ferocious carnivore—an exemplar of the warrior class. The Lion Man figure from Hohlenstein-Stadel cave in Germany (Figure 4.1, A), was recently dated to 40,000 YBP (contemporaneous with the extinction of *Homo neanderthalensis*). A dramatic and artistically mastered expression of the physical strength and bearing of a warlord or warrior king merged with a highly revered animal form. Unlike his African cousins, the male European cave lion does not have a distinctive mane. The numerous attributions of this sculpture as an early metaphysical or spiritual icon feel intuitively correct and fit well within the timing of our African self-awareness thesis.

And now we take a turn down the rabbit hole of political correctness. The Lion Man is not only referred to in many scientific references and research papers as the Lion Person, but has ardent supporters in the social sciences community who insist that, in fact, this is a figure of a Lion Woman! The physiological illogic of such a claim is dramatically confirmed by the large, rounded musculature at the joining of the upper arm and shoulder (deltoids) and the tapered musculature of the neck-to-shoulder transition (trapezius). We are familiar with this male body proportion in NFL tackles and WWF wrestlers of our day. The overall body form is clearly human, since the spinal cord arrives vertically to the base of the brain, not horizontally into the back of the brain, as is evident in Cave Lions, Figure 2.4, Detail C. The flat chest is, of course, unlike any portrayals of female form in the Middle Paleolithic.

Almost miraculously, given the precious few artifacts that have survived from this era, another ivory figure, made from a mammoth's tusk, was recovered in 2008 in a nearby cave, the Venus of Hohle Fels. Indisputably a woman and also dated to 40,000 YBP, it was found less than 200 miles from the site of the Lion Man. In Figure 4.1, B, we can easily compare these two contemporaneous depictions of the male and female form. I not only suggest that the Lion Man's gender is male but that these male and female sculptures are idealized from a male point of view.

This observation clashes with the characterization, in various scientific papers, of the many small "Venus" figurines from this period as being objects of worship or fertility goddesses. The fact that the Venus of Hohle Fels has no head other than a pierced stem for threading on a cord and is posed holding up her gigantic breasts and lifting her garment to expose an anatomy-defying oversized vagina centered over diminutive legs is one of the most clearly erotic and objectified depictions of a female in the archeological record or, for that matter, any record.

There is a remarkable similarity in the three sculpted Venus figures shown here, even though they range from approximately 40,000 YBP to 22,000 YBP. Eighteen thousand years pass and the figures all are nude, possess exaggerated breasts and buttocks and are made with genitalia exposed or wildly exaggerated. Although the Lion Man has clearly crafted facial features (confirming a skill that had existed over the entire 18,000 year period), and is head-upright, looking straight ahead, none of the female figures have facial features and all (if they have a head) are averting their gaze downward. The female figures range from 2.4 inches to 6.0 inches in height and are therefore easily transportable. By contrast, the male figure is 11.7 inches high and would certainly not fit in the palm of the hand. What could be a coherent explanation of these respective characteristics?

Ogi Ogas and Sai Gaddam, each with a Ph.D. in the field of cognitive neuroscience, have written the fearless book *A Billion Wicked Thoughts*, in which they tapped the "ultimate unobtrusive source of data," the Internet, to explore human desire from the perspective of a behavioral scientist.[17] Among their first observations from the Internet porn filtering software CYBERsitter, which blocks 2.5 million adult websites, is that 13 percent of the millions of searches they randomly collected were seeking erotic content and that 98 percent of all subscriptions to those pornography sites are made on credit cards with male names. Their research validated the "fundamental dichotomy" in the (heterosexual) sexual interest of men and women as reported by Catherine Salmon and Donald Symons in their book *Warrior Lovers*.[18] Salmon and Symons write that in the male fantasy realm

> of pornotopia, sex is sheer lust and physical gratification, devoid of
> courtship, commitment, durable relationships or mating effort...
> Porn videos contain minimal plot development, focusing instead on
> the sex acts themselves and emphasizing the display of female bodies,
> especially close-ups of faces...breasts, and genitals.

Ogas and Gaddam go on to observe that a woman's interest in sex is typically in the context of romance, a love story, and ultimately marrying the right man. Turning to the exact issue that we have under discussion, they say:

> Men's greater sex drive may be partially due to the fact that their

B
Venus of Hohle-Fels
2.4 in. high/ Mammoth ivory
Schelkingen, Germany
40,000 YBP

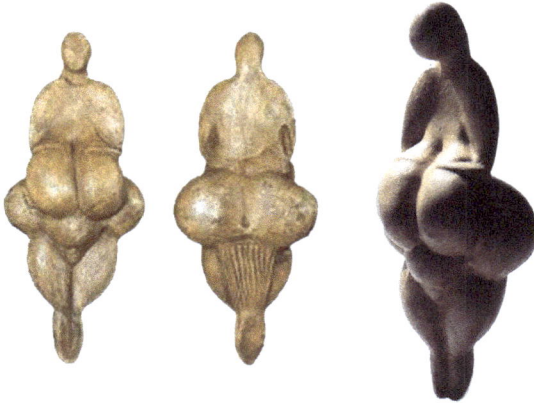

C
Venus de Lesphgue
6.0 in. high/ Mammoth ivory
Rideaux Cave, France
26,000-24,000 YBP

D
Venus of Willendorf
4.3 in. high/ limestone
Willendorf, Germany
24.000-22.000 YBP

A
Lion Man of Hohlenstein-Stadel
(top above and above)
11.7 in. high/ Mammoth ivory
Stadel Cave, Germany
40,000 YBP

Figure 4.1. Lion Man and Characteristic Feminine Form in the Paleolithic

sexual motivation pathways have more connection to the subcortical reward system than in women.

In other words, what you have long suspected is true; men's brains are designed to objectify females. This objectification of women extends deep into the mists of prehistory. The famous 26,000-year-old Venus of Willendorf statuette, hand-carved by a Cro-Magnon [*Homo sapiens*] in Paleolithic Germany, features GG-cup breasts and a hippopotami butt, but no face. The 40,000-year-old Venus of Hohle Fels boasts even more prodigious hips and mammaries—and titanic labia (and no head!)

Out of the one hundred highest-rated images on Fantasti.cc, twenty-three feature female anatomy without a face.[19]

The Venus figures are voluptuous and corpulent beyond any norm and certainly are in direct contrast to the super-skinny images of modern women in high-fashion magazines. The data of Ogas and Gaddam provide an answer:

For every search for a "skinny" girl, there are almost three searches for a "fat" girl![20]

The point is that both the idealized male and sexually objectified females are the embodiment of the Middle Paleolithic male point of view, a male presiding with a level of dominance not unlike the Janjaweed, Somali warlords in Darfur, or the Boko Haram militants of our day. The females are portrayed as prized possessions for sexual pleasure, and their physical forms communicate conspicuous consumption and wealth in the context of a violent world where starvation was a constant concern for most.

The Venus statuettes (particularly Willendorf and Hohle Fels) are so small that they fit perfectly in the palm of the hand. The legs on all three statuettes are diminished and come to a point—they were never intended to stand. The Lion Man is trim and muscular, and has solidly developed legs and feet, adequate and clearly intended to be displayed in a standing position.

The most ubiquitous hand-held object of the modern world is the iPhone, the painstakingly studied dimensions of which are intended to achieve the perfect ergonometric fit to the human hand. The dimensions of the original iPhone (still the most comfortable) were 4.5 inches high and 2.25 inches wide, a ratio of 2 to 1. The Venus of Willendorf, the average of the three statuettes shown, is virtually the same, 4.3 inches high and 2.25 inches wide! These bulbous and sexually exaggerated "handmaidens" are hard to imagine as anything other than personal, portable, erotic fantasy toys for fondling, totally devoid of identity or sacred purpose.

Homo sapiens at this point were still very close to the animalistic baseline of our species, and, for all their shortcomings, the best and the worst of our inheritance was forged there. Over time, we witness the falling-off of violence and the rise of gender equality and equal rights for all humans, in parallel with the rise of civilizations—not an illogical course of events.

MAPPING THE CONCEPT

Now that we have a perspective on the emergence of the right hemisphere and a general concept by which an infinite range of human personalities might arise, how can we visualize this new complexity of gender? One way to appreciate the diversity that flows from such a variation of the maleness/femaleness ratio is to complete a mapping of the neurological gender ratios, NGRs, across a representative population.

While this concept of one mind possessing two worldviews is difficult to express in the either/or mindset of our Western culture, it has strong precedents in Eastern culture.

One of the most familiar of these is the Chinese theory of Yin and Yang, which depicts the presence of two interdependent energies within a single circle. Paralleling our discussion, one side of the symbol, Yin, is characterized as Female (Moon-Passive-Conservation) and the other, Yang, as Male (Sun-Active-Destruction). They are described as "mutually dependent opposites" and symbolize the constant rising-and-falling cycle of these energies in life. Although the male/female characterizations are different, the dynamic concept, order, and visualization provide an excellent starting point for our mapping.

In Figure 4.2 we show the mapping of the individual mindset. A circle with two areas, one light and one dark, is used to represent the two sides of the mind. We can show the relative influence of one side or the other by increasing or decreasing the areas. For example, if the tai chi symbol for Yin-Yang (as seen below in the middle) is 50/50, or equal influence, by adding or subtracting the area of Yang (black) or Yin (white), the dynamic is illustrated:

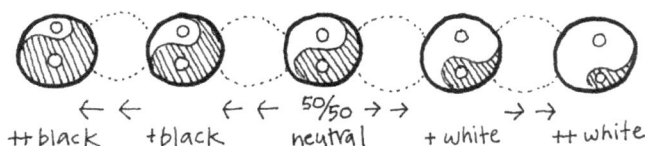

Figure 4.2. Mindsets: Variation of Right or Left Dominance

In our NGR representations in Figure 4.3, the genders are different colors (male = blue and female = red). The female symbol is shown on the right and the male on the left. Anatomical gender is represented by a small red circle, and the area of each circle represents its relative influence. In this case, they are all equal: 50/50.

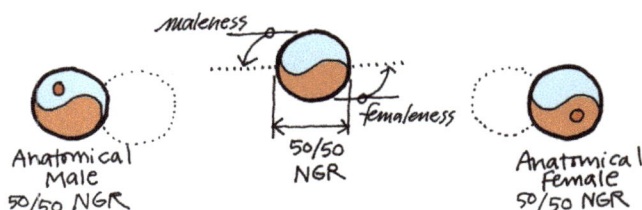

Figure 4.3. NGR Male and Female Symbols at 50/50 Equal Influence

The full range of neurological gender ratios is mapped in Figure 4.4. Anatomical males are on the left, anatomical females on the right. Row 1 shows the *Convergence* of anatomical and neurological genders from neutral to the maximum male on the left and from neutral to maximum female on the right. Row 2 shows the *Divergence* of anatomical and neurological genders from neutral to the transsexual *male-to-female* on the left and from neutral to transsexual *female-to-male* on the right. The counter gender effects are illustrated in Row 3 in greater detail (3a and 3b), which illustrates changing gender identity as transgender is approached.

We now have an abstract visualization of how NGRs could create the typical variability we observe across the human population and from generation to generation. These variations in neurological gender include the full array of sexual and non-sexual human predilections, talents, and perceptions which are bound up in each individual mindset in this process of diversification.

Two questions immediately come to mind:

1. What would be the biological pathway of such neurological diversity?
2. What evolutionary or adaptive advantage is achieved?

DNA/RNA: TWO DIMENSIONS OF HUMAN RESILIENCE

The modus operandi of DNA and RNA in the mind are the opposite of one another. DNA/natural selection is carrying forward surviving traits over time: two lineages (male and female) converging to the best fit of capabilities within the

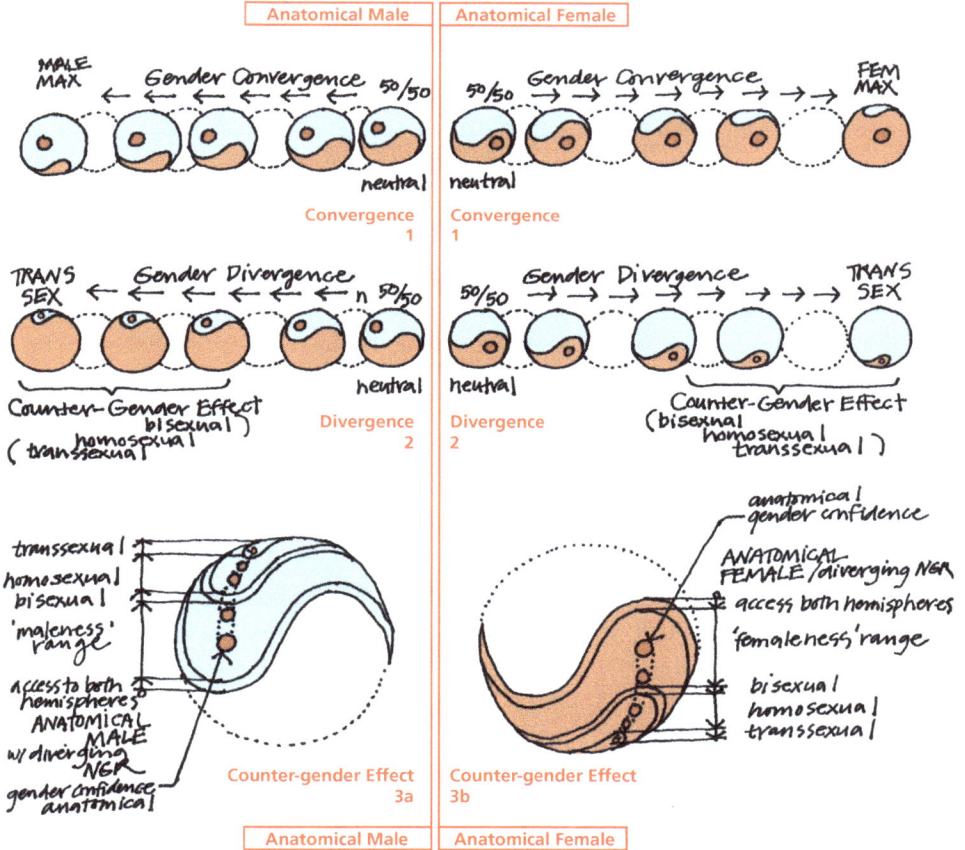

The Tai-ji symbol is used to depict the relative male/female composition of the neurological gender ratio: NGR. Male (blue), Female (red)

Figure 4.4. Linear Mapping of Neurological Gender Ratios

environment. RNA moves in the opposite direction by creating neurologically diverse perceptions, behaviors, and skill sets in each person, independent of the environment. RNA creates this diversity by *"expressing"* the evolved and underlying DNA differently in each individual.

In Figures 4.5 and 4.6, we illustrate how these two processes act through separate pathways and time frames to create high levels of resiliency in the human species. In Figure 4.5, DNA carries forward an advantageous trait (green), represented in 28 percent of the initial population. The disadvantageous traits, through early death, are eliminated within a thousand generations (20,000 years), and the green advantageous trait—let's say for night vision in *Homo neanderthalensis*—becomes dominant across the entire population.

Figure 4.6 illustrates the effect of RNA/NGR diversification in a catastrophic event such as a meteor strike or mega volcanic event followed by nuclear winter. The loss of life at impact will typically be no greater than the associated losses by mass starvation which follow. Generation 1 will experience the unforgiving post-apocalyptic survival test of mass starvation and unrestrained bloody competition for remnant resources. Reaching threshold events as severe as massacre and cannibalism, the total population in Generation 1 continues to be reduced by 72 percent (climate change and nuclear war can produce similar outcomes).

Most humans cannot or will not do what is required to persist in the face of such horrific circumstances. The portion of the population that will die in this catastrophic generation will likely be the males and females with the highest empathy/trust characteristics, the highest *femaleness* NGRs.

Survivors will be males and females of the most extreme *maleness* NGRs who are most able to prevail in these circumstances. These males and females, because of the RNA/NGR diversification effect, will continue to bear children in a wide range of *maleness* and *femaleness* NGRs. Even with long-term high-threat environments, with the continuing early death of those with high *femaleness*, RNA/NGR will persist in producing this diversity. Darwinian evolution through natural selection is too slow (thousands of years) to create the immediate responsiveness to crisis that is required for species survival in such an event.

It's important to note that Darwinian evolution cannot eliminate this continuing dimension of apparent "unfitness" to the surrounding environment, even if early death among the high-femaleness NGRs continues for hundreds of years, because natural selection is genetic, operating through DNA, while RNA/NGR is not genetic and operates independently.

In this manner, DNA and RNA work together to create a human species that is both resilient and "self-righting," much as a sailboat spills more air from its sails by tipping lower in heavy storm winds and then rises higher to catch more wind as the storm recedes. Once a stable and durable defensive perimeter is reestablished, the less aggressive and more creative femaleness males and females, no longer at risk, will populate and begin to generate competitive advantages within the resurrection of a neurologically diverse population, as seen in Generation 5 in Figure 4.6.

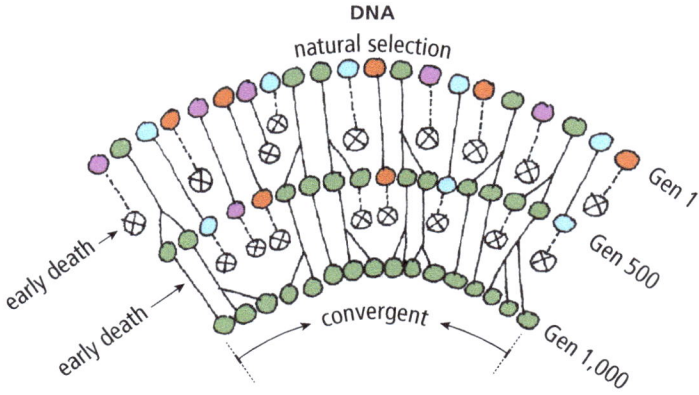

Figure 4.5. DNA: Neurological Trait Achieving Full Effect in 1,000 Generations (20,000 years)

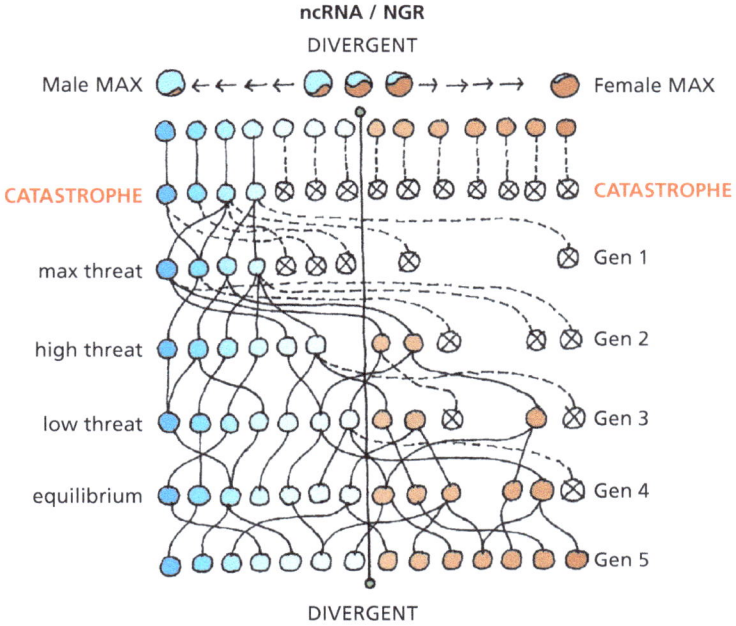

Figure 4.6. RNA/NGR: Neurological Gender Diversity Restored in 5 Generations (100 years)

ALL POSSIBLE FUTURES

The catastrophe described here is only one of the many possible futures for which we have been prepared. Each one of us is evolution's bet on fitness in the face of future eventualities. While those circumstances are not knowable, it doesn't take more than three examples to demonstrate that very different mindsets and skill sets will be needed to survive and eventually thrive.

We know that there are two pathways forward on the sustainable use of Earth's resources: continued unsustainable consumption or a sustainable balance of consumption and restoration of natural systems. We are in the midst of an increasingly polarized debate on which of those two roads to take.

The geological history of the planet provides us with a third scenario.
We know that there have been catastrophic events over the millennia that have extinguished life on a vast scale, one of which eliminated the largest and most fearsome life forms to have walked the Earth: the dinosaurs.

Mega-volcano events and/or meteor strikes created multiyear periods of nuclear winter, limiting crops worldwide, forcing drastic resets for all life forms (population crashes, extinctions, and bottlenecks), and challenging the "fortunate" survivors to their limits. (A full nuclear exchange, like the one averted between the United States and the U.S.S.R. during the cold war contention over Cuba would, in all probability, do the same.)

In each of the three following scenarios, natural selection rewards very different mindsets and skill sets. The human population contains individuals predisposed to succeed in each scenario.

1. **Non-Sustainable Future**
   A world characterized by increasing conflict over limited resources, migrations and border incursions by the starving and disadvantaged, and economic collapse. A highly militarized environment follows.
   Culture = pragmatic, militaristic and bureaucratic, low trust/high doubt, low creativity.

2. **Massive Meteor Strike, Volcano Cluster, Nuclear Warfare**
   A world characterized by several years of cannibalizing the remnants of the pre-event world, followed by population crash, violence, and chaos.
   Culture = anarchic, power-based, no trust/all doubt, survivalist, low to nonexistent creativity.

3. **Sustainable Future**

   Highly concentrated and efficient urban centers, greatly reduced rural and suburban development, balanced population growth versus ecosystem services, and a high valuation of innovation and creativity. A world characterized by minimal violence and high equality.

   Culture = high creativity, arts, and innovation with collaborative endeavor, high empathy, high equality, high trust/low doubt.

In the worst scenario (#2), it appears that the segment of the population we consider least fit (sociopaths and psychopaths) could initially rise to power, being able to kill without hesitation or remorse. Since this segment of the population holds steady at 4–5 percent, think of it as the maintained "ace in the hole" in order to save the species in the darkest of futures.[21]

No matter how severe the population crash envisioned in scenario 2, and no matter if it is five or fifty generations during which the empathetic and creative humans are selected out, each new generation will bring their rebirth. When high levels of innovation, empathy, and creativity are valued (in an extended period of domestic security), those humans will rise to the fore, and the swing back to a stable and more equitable civilization will continue. It is important to underline that a precondition will have to be met: territory will have to be held against attack to provide sustenance as well as the time and space required for birth, nurture, and child care through maturity.

THE NORMAL BRAIN

DNA and natural selection concentrate viable and proven ancestral male and female traits in the current generation. A useful analogy is to think of DNA as creating male and female books of "best practice" for survival as the hemispheres of the brain. The formative fetal brain contains 100 percent of all these possible human outcomes, but we know that a single person cannot utilize or be all of those contrasting and, in many cases, mutually exclusive traits.

Nor would a single person be totally influenced by the left or right hemisphere, or there would be no debate (deliberative thought) between opposite minds. This is where RNA comes in as the final editor and composer of the mind, creating a unique and workable human-scale subset of these possible traits.

The strategic distribution of these advantages creates segments of the population that are highly prepared to survive in worst-case scenarios and other segments that are highly prepared to thrive in best-case scenarios. In other words, the species as a whole is more secure in times of great risk and times of great opportunity because there are those who are, effectively, living in different worlds, who see and experience the same events in different ways. The most resilient population is achieved when the ideal perception of events—and therefore the ideal response—for every eventuality exists somewhere within the population.

For instance, in the United States Civil War, a little-known leather-goods shopkeeper from Ohio, Ulysses S. Grant, who had earlier resigned from his military career, rose to head the U.S. Army over the many politically connected luminaries and top-of-class graduates at West Point (he finished twenty-first in a class of thirty-nine). Grant, however, had the perseverance and sheer guts to drive his superior manpower and logistical support to unambiguous and often bloody victory while his superiors had failed that test for years. Unsuccessful in civilian business pursuits before being called out of military retirement (and labeled a hothead and an alcoholic by some), Grant embodies a familiar story: the traits of the warrior not being transferrable to peaceful civilian pursuits and yet existing within our population and ready to serve in times of crisis.

In a more contemporary example of extreme variance, we may view the survivalist with a deep bunker and six months of provisions in the mountains of Montana as a right-wing nut, but on the other hand, if Earth is struck in a series of impacts like the Comet Shoemaker–Levy 9 that struck Jupiter in July 1994, he may be among the few to survive. A resilient population viewed at any point in time is going to have a considerable number of people who are "way out in left field," not suited to the environment at hand.

In contrast to our vast diversity, a population of uniformly "normal" brains with a single worldview would all respond similarly or in unison to threats and opportunities. It would be just a matter of time before a unanimous, fatal choice was made. This is the opposite of a resilient population.

THE BIG PICTURE

Darwinian evolution and the concept of "survival of the fittest" through a process that converges, over time, to an ideal human form was an irresistibly elegant explanation that was consistent with observations in nature, and it has ultimately proved to be an amazingly productive avenue of scientific inquiry. The power of the idea was such that other dimensions of our reality that did not fit this model were marginalized or characterized as abnormalities and the result of "environmental" factors. By extension, three general conclusions followed:

1. There is a single universal human nature (independent of gender).
2. There will emerge, through scientific inquiry, a key biological

mechanism for our convergent evolution to the highest level of fitness, and that key is within DNA.

3. This high level of fitness is embodied in our physical body and brain and is achieved across the population with a relatively small number of exceptions and abnormalities.

The assertion under discussion here is that there are two biological mechanisms of human-species resilience at work on the human mindset, which is composed of two human natures. DNA/natural selection is genetic and imparts resilience at the scale of the individual through adaptation over time, whereas RNA/ neurological diversity is not genetic, and imparts resilience at the scale of the species immediately in the current generation through unique interpretations of the baseline DNA mindset. This process creates a higher level of complexity and resilience of humankind and is reflected in the high level of variability of human natures observed across the population. By extension, three general conclusions follow:

1. There are two formative life experiences, or human natures—one male and one female—which, through DNA/natural selection, are carried forward for expression in the left and right hemispheres of the brain, respectively.

2. Human fitness, or resilience, results from two separate dynamics. One is convergent: reducing diversity (DNA/natural selection) and one is divergent: increasing variability (RNA/neurological diversity).

3. RNA/neurological diversity creates unique ratios or expressions of the two human natures collected via DNA/natural selection; best expressed as Neurological Gender Ratios: NGRs. The resulting spectrum of perception, and behavior, creates a level of diversity and resilience across the population that DNA alone could never achieve.

DNA concentrates fitness in the individual for adaption to a specific environment; RNA creates a separate layer of *nonadaptive fitness* in the species (neurological diversity) for many possible environments. The selected DNA in our current generation is a function of past events over vast periods of time. RNA, by contrast, imparts at birth a pattern of neurological diversity across the species as a whole that enhances the resilience of humankind in the face of unknown and future events. Acting together, these separate and, in a sense, opposite phenomena have the effect of maximizing the resilience of the human species.

THE RNA/NGR 'DASHBOARD'

If we give RNA a mission that involves only three variables in each birth, anatomical gender (male or female) and variation in the degree of influence between the hemispheres of the brain (maleness *and* femaleness), we should be able to map out a simplified spectrum of variation. In Figure 4.7, I have created such an interpretation. Even with only three variables, we observe a wide-ranging outcome by integrating anatomical and neurological gender. Most importantly, even at this level of abstraction, it reflects a population that closely resembles our own.

For clarity, this continuum is organized in a clockwise flow with anatomical females on the right sweeping downward (maximum *femaleness* to minimum) and anatomical males on the left sweeping upward (maximum *maleness* to minimum). The neutral balance of gender influence occurs in the middle, at the clock positions of 3:00 and 6:00. The transgender effect of counter-gender identity creates the link of continuity between the male and female sides and occurs at 6:00 (female to male) and 12:00 (male to female).

Gender expression begins with maximum *femaleness* in anatomical females at 12:00 and proceeds clockwise with increasing proportions of *maleness* until past 5:00 where there are counter-anatomical gender expressions observed and finally male identity expression (transgender) occurring just before 6:00. From 6:00, sweeping upward, the anatomical male side begins with maximum *maleness* expression and proceeds clockwise with increasing proportions of *femaleness* until past 11:00 where there are counter-anatomical gender expressions observed and finally female identity expression (transgender) occurring just before 12:00. In this coherent manner, all variables are expressed and the skills, talents and predispositions associated with each of the Neurological Gender Ratios are imparted.

Such a spectrum also has the advantage of being a very spare instruction or, in math terms, a simple algorithm. We may critique the transgender effect (surgically reversible in a modern context) but such statistically small outcomes seem to be swept up in the massively beneficial large scale expressions. For this reason transgender will naturally reoccur in subsequent generations because it is bound up in an overall RNA-neurological gender process which is independent of the DNA-natural selection process.

This visualization is, of course, a vastly simplified version of a much more complex process with many other variables. However, overlaying this baseline with self-awareness creates a revealing signature which has deeply imprinted itself on the social and cultural expressions of the Third Millennium Mind, the subject of Chapter 5.

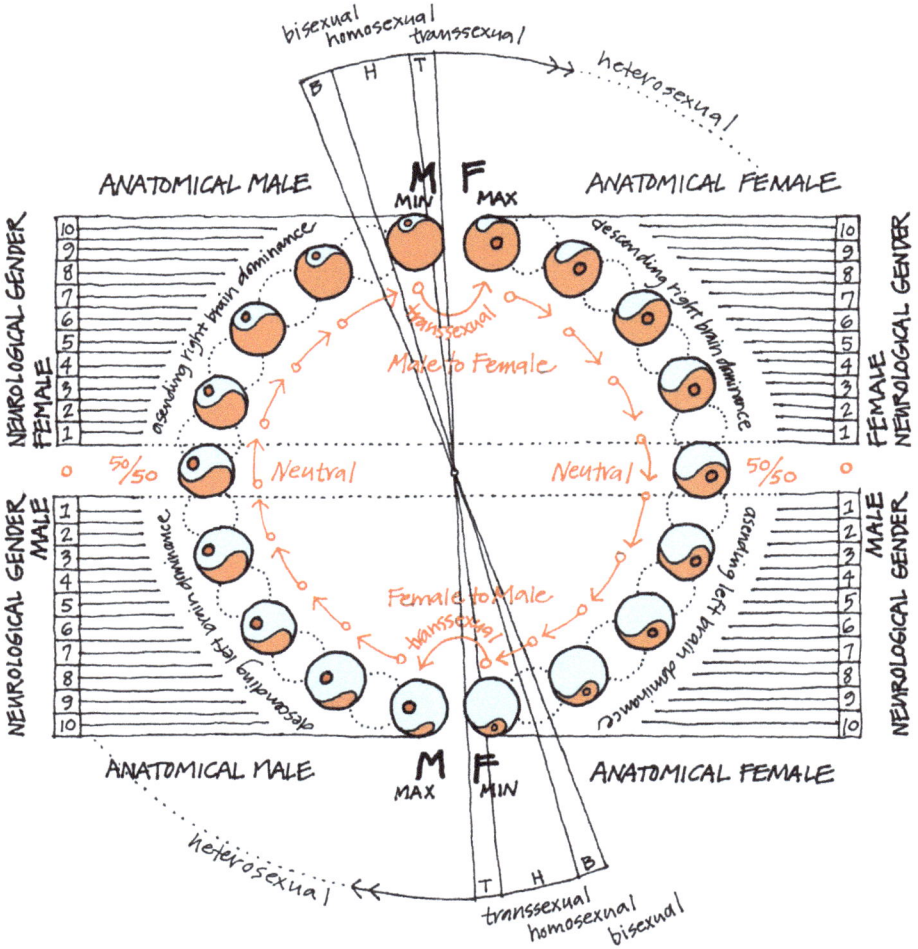

Figure 4.7. RNA/NGR: A Spectrum of Diverse Mindsets

Author's note on Figure 4.7: If we could look over the shoulder of RNA and see the mapping of neurological diversity across the population of anatomical males and females, so that the range of human personalities we encounter every day are achieved, this is an abstraction of how such a spectrum might be ordered.

CHAPTER FIVE

The Third-Millennium Mind

IN MUCH THE SAME WAY THAT SCIENTISTS SCAN THE BACKGROUND radiation of the universe (cosmic microwave background, or CMB) as a way to understand our starting point, the Big Bang,[1] it seems logical that we should search for the traces of our beginnings here on Earth.

Our premise throughout has been that we began in an animalistic state characterized by male dominance, which included interpersonal violence at the scale of the family and intergroup violence in the form of raiding and the murder of males and most females, with the exception of the rape and kidnapping of young women. Biologically, we have asserted that this ancestral male life experience or *male survival pathway* informed the brain's left hemisphere (maleness). Females, in addition to the instances of rape and kidnapping, followed the biological imperative of incest avoidance by migrating to other bands and, in both instances, gaining acceptance. We have asserted that this ancestral life experience or *female survival pathway* informed the characteristics of the right hemisphere (femaleness).

When we look around the globe, do we see the lingering traces of such a beginning? The answer is yes, and one wonders how anything could be more self-evident. Still existing at this primitive bottom rung are the Janjaweed militia, whose raiding of settled communities in Darfur is characterized by the most extreme levels of murder and torture of males and sexual violence toward women.[2] According to the Genocide Intervention Fund, "The Janjaweed militia use sexual assault to cleanse Darfur of its ethnically distinct African population—eugenics through rape." [3]

Rape, gang rape, and sexual slavery perpetrated by the Hutu militiamen on women primarily of Tutsi ethnicity were a central aspect of the Rwandan Genocide of 1996. More recently, in 2014, the UN Mission in Iraq and Amnesty International confirmed that hundreds of Iraqi Yazidi women have been sexually abused and taken as slaves by ISIL. These are compelling examples of such *background radiation*, revealing the stubbornness of the nature of our animalistic beginnings and the latent capacity to revert to such a state. This radiation continues to exist and flare up in our modern world, but, thankfully, it now occurs as an exception to civilization and the general rule of law.

As we move up the list of less violent but still male-dominant cultures, we observe an incremental decrease in violence as in the cases of Saudi Arabia, Pakistan, and India. Once we arrive at the highest levels of justice, the lowest levels of corruption, and the lowest levels of gun ownership, we encounter the global minimums of violence and, consistent with our premise, the maximums of gender equity. Norway, New Zealand, and Austria are such countries.[4] Needless to say, national and domestic security, directly or indirectly, must exist as a precondition for such a society.

There are several methods of achieving the required nonviolent setting. One of these is military dominance, or equilibrium with adversaries, or to gain, through strategic alliances, an assurance of military support to discourage an attack. At the other end of the spectrum, one may choose simply to retreat from violence by moving to a remote, uncontested, and probably subsistence life-style such as can be observed in the hunter-gatherer communities of the Inuit and isolated Amazonian tribes. In either method, the society achieves the degree of separation from violence and daily threat that tends to free up the right-brain, femaleness qualities of altruism, empathy, and equity between men and women.

A pattern emerges as we move from the global extremes of most violent to least violent, from most male-dominated to the most equitable (there are no female-dominated examples): a pattern of moving from the realm of the left hemisphere to the realm of the right hemisphere. Among nations, we also observe a chronological characteristic such that left-hemisphere dominance tends to be ancient and right-hemisphere primacy tends to be modern. This reflects the good news: all nations are, in the longest of long runs, upwardly evolving and moving from left- to right-hemisphere emphasis.

JUNE 2005

The concept sketch in Figure 5.1 was completed in the summer of 2005 as I was trying to organize my thoughts on how maleness/femaleness of the hemispheres of the brain might be expressed both internally and in a modern cultural context. The wandering line down the middle of the diagram represents the wiring of the two hemispheres of the brain. The original sketch is in blue ink; the red

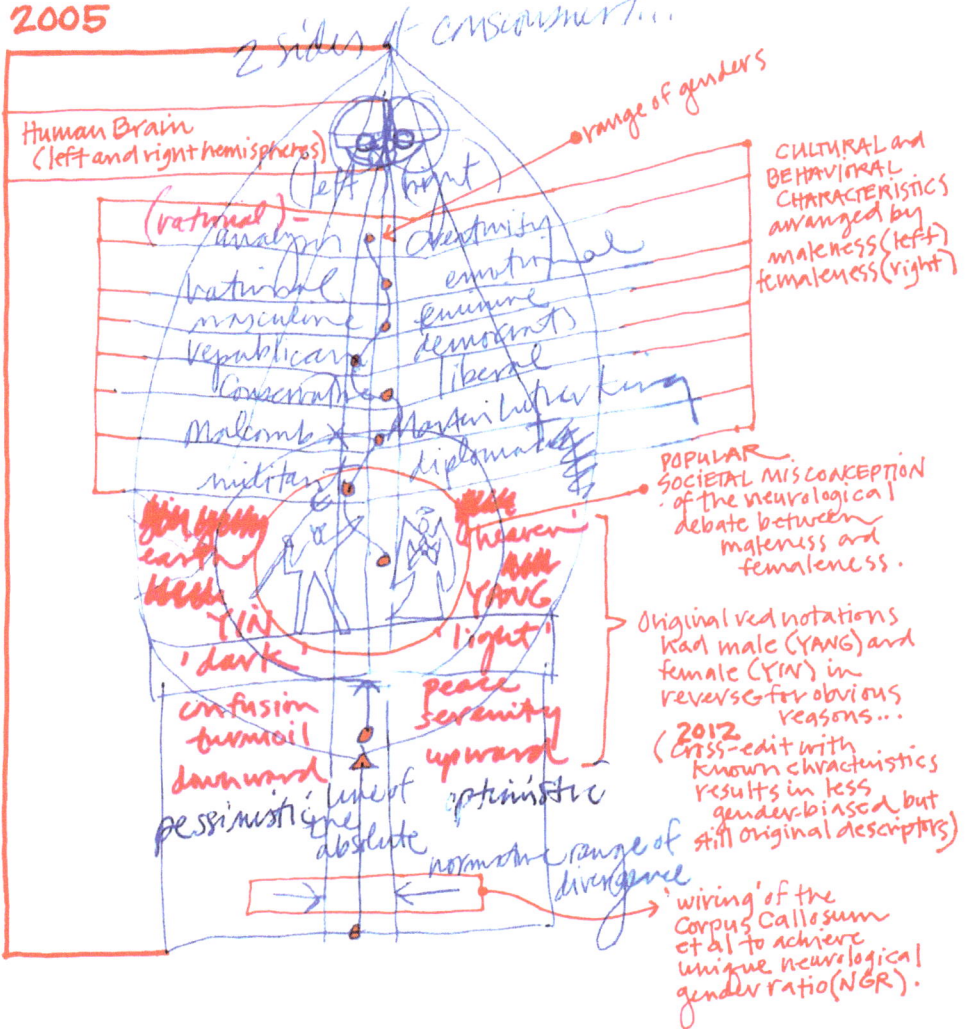

Figure 5.1. First Conceptual Sketch of Neurological Gender Ratio (NGR)

clarification notes were added in 2012 at the time of a first (unpublished) version of this book.

The impetus for this jotting on the back of an envelope was running into E. O. Wilson in the sitting lounge of the Harvard Faculty Club. I was so energized by our brief conversation that I was inspired to sketch out my own concepts of a biological basis for the variability of human mindsets (with a few misplacements and reversals of characteristics). My first follow-up effort to get words on a page, in 2006, never got past an initial thirty-page treatment, which was set aside under the day-to-day demands of running an architectural practice in New York City. As imperfect as it may be, my initial sketch captures the purpose of this chapter: the consideration of how such a neurological gender framework of the brain would be expressed in the culture of the third millennium.

For comparison, I have included the more detailed 2012 version of the same exercise, on the opposite page, which incorporates a "wiring diagram" for the variations in emphasis of each subject in relation to each hemisphere of the brain—a suggestion of the complexity that is quickly achieved in the articulation of our "two minds." Both sketches are speculative and overly simplified, but they begin to show how the variability of maleness and femaleness characteristics and predilections, as illustrated in the RNA Dashboard in Chapter 4, can quickly create humanity's wide range of outcomes.

These two-dimensional depictions of geographical location in the left or right hemisphere are multiplied in complexity when we remember that they are occurring simultaneously—often not even on a conscious level, as part of our internal conversation. In this way, we experience the richness of our being, with influences flowing seamlessly from both sides of the brain as well as from the environment we are in at the moment. The intention here is to convey the complex beauty of maleness (protect and defend) and femaleness (trust and creativity) as two halves of a universal consciousness, an internal deliberative basis for judgment. Our ancestors, over the 100,000 years of humanity, have contributed to that mindset by making sound survival decisions through at least the age of reproduction, thereby preserving the unbroken line of this inheritance.

To sum up for clarity in this key chapter, when I speak of the characteristics of an individual, I am speaking about his or her neurological gender ratio (NGR) of maleness/femaleness, which is referenced as being dominant on one of the two sides or in equal balance. It is important to remember that no one is 100 percent of either extreme or there would be no debate between our "opposite" minds. We all have an anger and territoriality response and we all possess feelings of empathy and justice. Admittedly, there is a chasm of difference in the relative intensity of such predispositions in any individual, but it's not as though either side is totally absent, even in the extreme emotional desert of the sociopath and psychopath.

When we consider the accomplishments that have flowed from our diverse population in the areas of diplomacy, governance, science, architecture, global

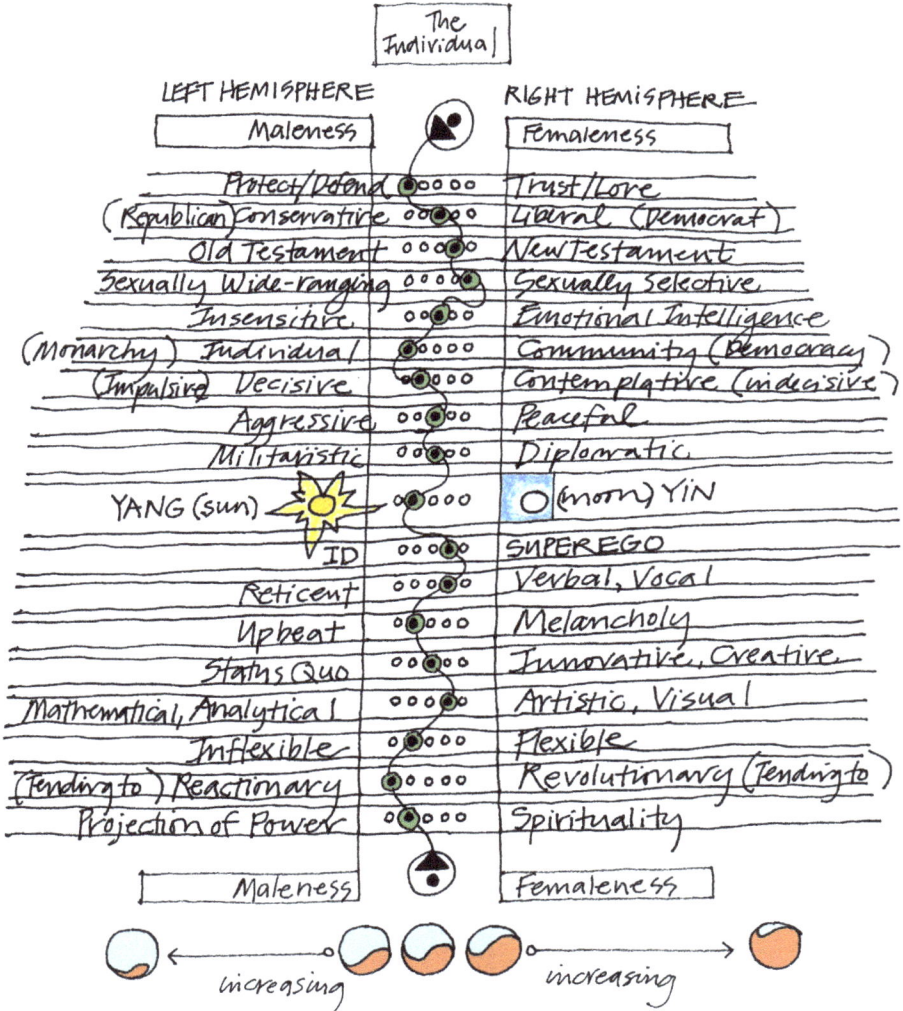

Figure 5.2. Elaborated Mapping of Cultural Behavior (2012)

communication, design, theater, television, the Internet, fashion, art, music, and other creative endeavors, we are dazzled by the array of colors displayed. Although we don't think of these as weapons of global competitive advantage, they are a national treasury of creative expression and intellectual equity that can have as much to do with the long-term viability and competiveness of a nation as a stockpile of weapons.

THE PEACOCK'S TAIL

One metaphor for a society that could shape its governance so as to take maximum advantage of each individual's creative and intellectual potential and yet maintain the necessary national defense would be the extravagance of the peacock's tail. The peacock's elaborate and colorful fan of feathers has puzzled researches for years because it seems to be contrary to survival of the fittest. Why risk disclosing your presence to predators with such a massive and visible display? Since it is obvious that the peacock has not had a problem with survival, it is clear that in the competition for mates, the species has found that key balance between the pragmatic need to survive with the maximum expression of beauty to guarantee reproduction. The creative diversity found in a fully representational democracy achieves a similar richness and thereby achieves a global strategic advantage in that it is by far the most beautiful, the most attractive to others.

To the extent that a government, for reasons of religious conformity or security or status quo, or simply out of fear of losing control, marginalizes all or part of its population by denying participation in an open society and self-governance, it has crippled its national, innovative, and culturally competitive assets. Cultural advantage becomes a determinant strategic advantage in a stable state of no war/low violence as currently exists among the major global economic powers. Instead of raiding neighboring tribes to kidnap reproductive capital, the modern peaceful version is the representative democracy that attracts the best and brightest (intellectual capital) as well as investment (financial capital) from the other nations of the world.

An explicit example of this effect, on a smaller scale, can be seen in the early competitive model and strategies of Apple versus Microsoft. The early operating system of Apple was more elegant and accomplished more with less code than Microsoft. However, Apple jealously guarded its code. In contrast, Microsoft decided to make its code accessible to software developers in the open market, allowing them to create their own profit centers. An explosion of PC-only creative applications soon swept Microsoft into a dominating position in the marketplace for computers, despite the fact that its operating system was less elegant and more patched together. The lesson was not lost on Apple, which, in its recent resurgence, has opened up its code for the creation of thousands of signature apps. While this seems at first to be a similar open marketplace of ideas, Apple's interpretation of "open" requires a share of profits from the developers, who must agree to sell through Apple's App Store. Both surges of innovation were the result of opening up and leveraging the creative potential existing in the broader population and marketplace.

Transparency, accessibility, and open system design are the essence of a democracy that, at the scale of a nation, can accomplish this same corporate example of optimization of potential. It is important to point out that I am not putting the United States forward as the idealized model of democracy, in fact, none of the world's democracies is fully realized at this time. A stunning reminder of just how slow the upward evolution of democracy has been is the fact that women did not achieve full voting equality in the United States until 1920, in England until 1928, and in France until 1944!

Nonetheless, the metrics of success are obvious in the thirty-four democracies that are members of the Organization for Economic Co-operation and Development (OECD). Representing only 18 percent of the world's population, these countries account for 63 percent of world GDP and 75 percent of world trade. It is important to note that humanistic and intellectual capital metrics are reflected in similar proportions, all of which supports the recognition of democracy as the current biological consensus model for governance of the human species. In fact, this is where we have been headed in a biological sense from the first day of self-awareness.

In order for a nation to reach its full potential, the recognition and full integration of all ranges of diversity of its citizens is essential. The closer a nation approaches that ideal, the closer it is to behaving as nature's paradigm of social order, the superorganism—an "open system" that is in constant formation and reformation to best address threats and opportunities in the surrounding environment. One example in nature is the honeybee colony, a superorganism selected by the ancient Egyptians to symbolize one of their early kingdoms—an exemplar of industriousness and harmony. To the extent this ideal of social order is achieved, democracy functions as an ecological entity, that is to say, it functions in an upwardly evolving evolutionary manner.

We now consider how we have transitioned to democracy and the complexity of our modern world through evolutionary and biological stages, beginning with humankind's first dawning of self-awareness 100,000 YBP in Africa.

TRANSITIONS TO THE SUPERORGANISM

There have been three transitions in humanity's rise from the scale of the band to the scale of the superorganism. Each step has been a revolution in that each has been accomplished worldwide and, primarily, by violent overthrow. The stage 1 transition, fueled by the emergence of self-awareness, led to the extinction of the remaining genus *Homo*. Stage 2 was the forcible replacement of the hunter-gatherer culture by the elite model of grounded (agrarian) culture, again at a global scale. Stage 3 is democracy, which is replacing the elite model and has broken through to the modern era and the threshold of the superorganism. Although stage 3, currently in progress, may seem disorderly and slow, it is by far the fastest of the three transitions.

The violence of stage 3 that attended the American and the French Revolutions followed the familiar bloody patterns of stages 1 and 2; however, the sequential transition to democracy and away from the role of the monarchy by Great Britain proved to be less violent. The remaining democratic transitions occurring across the modern world have been, and increasingly will be, facilitated by the transparency and immediacy of digital communication in our age of the global village.

These large-scale transitions in size and complexity, shown in Table 5.1, support a biological provenance since each stage is sequentially more efficient, with a rise in population density and an increasing specialization of minds. In other words, there is convergence on the fundamental characteristics of the superorganism. The chart organizes these stages by chronology, impetus, and most distinctive characteristics.

The recent claims by the nondemocratic regimes of Russia and China to be nations that value freedom and justice, and their increasingly public (albeit unconvincing) displays to support such claims, constitute a grudging recognition of the power of democracy and a halting step in that direction. A refreshing pragmatism is captured in the apocryphal response of a Chinese official who, when pressed on the subject of socialist democracy, capitalism, and free markets in China, said: "We're going to find what works and then we're going to call it Socialism."

Beginning as a limited concept in the Greek city-state of Athens 2,520 YBP and broadened to function within a full societal context by the philosophers of the Enlightenment, democracy only reached full structural form with the adoption of the U.S. Constitution (1788) and the French Declaration of the Rights of Man and the Citizen (1789). Representative democracy is the first form of human governance to so closely emulate biological and evolutionary principles. Amazingly, the laws of nature were cited throughout the history of democracy, notably by Aristotle, Plato, John Locke, and John Adams.

Why do we say that democracy is the first form of governance to function on ecological or evolutionary principles? Because the reference model in nature is the superorganism, which quickly perceives threats in the environment and takes collective action to assure survival. For all the success of the elite-centered civilizations headed by pharaohs, monarchs, popes, or even a central committee, they nonetheless all suffer from inflexibility and decay at the center. Isolation from the governed grows over time, which undercuts the ability to perceive and respond to ever-changing surroundings.

How can a government grapple with the mega-question of trust at the scale

Table 5.1. Evolutionary Transitions from Band to Superorganism

Stage 1 Transition	Nature of Change	Biological/Cognitive	Advent of *Homo sapiens* (destruction of the rest of the genus *Homo*)
	Transition	Animal to Human	
	Time Frame	7,500,000-100,000 YBP	
	Duration	7,400,000 years	
	Indicator	Self-awareness	
Stage 2 Transition	Nature of Change	Secure Settlement and Controlled Food Source	Advent of Civilization Super Organism (destruction of Hunter-Gatherers)
	Transition	Hunter/Gatherer to Elite Grounded Communities	
	Time Frame	100,000 YBP-6,000 YBP	
	Duration	94,000 years	
	Indicator	Agriculture/Fortification	
Stage 3 Transition	Nature of Change	Cultural/Philosophical	Advent of Human Super Organism (destruction of Elite Control)
	Transition	Subject to Citizen	
	Time Frame	2,520 YBP-Present	
	Duration	On-going	
	Indicator	Democracy	

of a nation? At the scale of the individual the question is whom can I trust in combat, or marriage, or to handle my money, but at the nation scale, it becomes whom can I trust in war as an ally, or to honor a peace accord, or to partner on a space mission? These are the do-or-die questions of a nation's continued existence. The mega-question of trust—a contextual question, sometimes yes, sometimes no—is therefore the ultimate question of survival at the scale of the individual and of the nation.

Even the most radical young and well-intended leadership acting with full support of the populace at the moment of victory will eventually become out of touch and weighted down by accumulated self-interest. This is why no planned economy or state has ever survived. No matter how well-meaning or how brutal—Pol Pot and Idi Amin come to mind—they fail. Democracy forces the constant refreshing and recasting of leadership within the ongoing feedback of the governed, who are at the interface with the environment.

Forms of democratic governance are informed (and reformed) with new minds and an aggressive debate on the same fundamental subjects of survival and trust that go on in our bihemispheric brain. At the scale of the nation, a democracy creates a congress that is the superorganism version of the congress of our minds, and the opposing forces in our governance are the same liberal/ femaleness and conservative/maleness views of the world that are present in the hemispheres of our brains. This self-critical process, this existential debate, is the wellspring of evolutionary advantage and thereby the survival and prosperity of the nation.

The important point is that when a fully representative group of minds is assembled from the populace of a nation, access is gained to humanity's full base of knowledge, which does not—and cannot—exist in any one individual; this is the power of a democracy. The fact that there are consistently close votes and tight electoral outcomes on the liberal or conservative side of an argument reflects the uniform biological distribution of these mindsets throughout the population.

The swings back and forth in liberal or conservative administrations are based on what is perceived at that time to be most critical by the electorate (the ones out on the edge of the real world). The key segment of the population in these swings is the neutral or 50/50 NGR group that constitutes the independent voters. They are balanced biologically between maleness and femaleness and therefore are the most able to consider the world from either point of view. The outcomes of the popular vote in the United States presidential elections point to a specific percentage of the overall population that is effectively 50/50 NGR. The two largest landslide elections in U.S. history were Johnson/Goldwater in 1964 and Nixon/ McGovern in 1972. One was a liberal victory and the other conservative, each 60-40. Given that these both reflect the greatest margin of victory ever achieved, it would appear that 80 percent of the population (40 percent conservative and 40 percent liberal) is almost immovable in its political worldview and that the more flexible 50/50 NGR grouping is no more than 20 percent of the population.

Now we move on to more challenging questions. If the two hemispheres of the brain embody ways of being in the world (simplified as liberal/femaleness and conservative/maleness), aren't they essentially a choice between good and evil? After all, peace, understanding, and love are good, and war, violence, and hate are evil. Aleksandr Solzhenitsyn's novel *The Gulag Archipelago* beautifully frames an answer:

> If only there were evil people somewhere insidiously committing evil deeds, and it were necessary only to separate them from the rest of us and destroy them. But the line dividing good and evil

cuts through the heart of every human being. And who is willing to destroy a piece of his own heart? [5]

Another characterization of these forces is captured in Davis Grubb's novel *The Night of the Hunter*, the story of a preacher-turned-serial killer, Harry Powers, portrayed by Robert Mitchum in the 1955 movie of the same name.[6] Memorably, Reverend Powers had tattooed on the four knuckles of one hand the word HATE and on the other the word LOVE.

Finally, in the 1919 case Schenck *v. United States*, we have the words of Supreme Court Justice Oliver Wendell Holmes Jr.: "The character of every act (good or evil) depends on the circumstances in which it is done."

Among the three examples presented, Solzhenitsyn's words strike home with an observation that we all know to be true: evil impulses and ideas do emerge from our inner landscape, and each one of us has a capability for both good and evil. The expression of these opposites is powerfully rendered in the "knuckle" iconography of Reverend Powers, in that LOVE requires the deepest form of unconditional trust, and HATE is a total absence of trust.

However, it is the words of Justice Holmes that are most relevant in an evolutionary sense. They carry a core truth, the realization that the meaning of any act grows from its context.

If we in the United States celebrate our part in the ending of World War II, then we must acknowledge that our actions are only justifiable against the larger and more frightening prospect of a victorious Nazi Germany or we would not be able to defend the bombing of Dresden and other major German civilian populations. Above all, the explosions of atomic bombs over the towns of Hiroshima and Nagasaki that were deemed essential by our government for final and complete victory in that war still haunt the collective conscience of many. When is the killing and wounding of hundreds of thousands of men, women and children—the majority of whom are not military combatants—*not* an evil act?

As reluctant as we may be to admit it, such wrenching decisions are a continuation of evolution's imperative of competition carried out in the concentrated and mechanized form of war. We call it—at least in this case—justifiable. Within our legal framework, as well as our religious framework, our actions in World War II were heralded and, due to the actions of our "greatest generation" in the aftermath of the war (including the Berlin airlift, the Marshall Plan, and the Reconstruction of Japan) the outcome is viewed, overall, in a positive light by the global community.

Good and evil exist as clear boundaries of conduct only within human laws and adjudication, developed as an extension of a universal moral sensibility. This is an internal framework of good and evil within our national borders, while in terms of external or global conduct, there have always been rewards for those who prevail over their foreign adversaries. The good news in modern times is that a general equilibrium of force has been struck between the world's major powers. Mutual economic interest and the transparency of instantaneous global communications have helped to expand and institutionalize this low-threat environment.

When we say that the complementary but opposing neurological forces of the right hemisphere (female/liberal nature) and the left hemisphere (male/conservative nature) are interdependent and that neither is good or evil, we have framed a striking parallel to the Chinese Taoist philosophy of Yin-Yang—the two life forces that symbolize the interlocking dynamic of life. If we overlay this Eastern philosophy as a fundamental point of view over a format of argument and counter-argument of a Western democracy such as ours, we recognize the universal quality of interdependence and the inherent mutual benefit to be gained. This existential debate is the wellspring of evolutionary advantage and, thereby, the survival and prosperity of a nation.

From the YIN (right hemisphere) side of the argument:

Liberal (Yin) Declaration: Conservatives are "hawks," always seeing the worst in others and ready to start a conflict at the slightest provocation. Proponents of a strong national defense, they seem to think that the only role of government is to protect and defend the nation (the status quo). Constantly overestimating threats, they are fiercely resistant to diplomatic efforts and peace initiatives. As a nation, we should be doing the exact opposite; we should take risks in the pursuit of world peace, eliminate nuclear weapons, and reinvest those resources to achieve sustainability and end world hunger.

Conservative (Yang) Counter Argument: The United States has been blessed with geographical isolation, which has contributed to the very small proportion of our population that has been killed in foreign conflict. The remoteness of wars over time has allowed us to forget that, in the beginning, Americans killed British soldiers, Hessian mercenaries, Native Americans, Spaniards, and Mexicans in the violent process of taking the territory to form our nation.

If we as Americans consider ourselves blessed in any way, we need to be thankful that our forefathers had the initiative, courage, and perseverance to take and hold territory, and the foresight to press our advantages by purchasing or otherwise acquiring the rest of our nation.

The first global conquest by *Homo sapiens* was accomplished through violent force and the willingness to kill or be killed. Exactly that bloody level of resolve has been present in the creation of essentially all nations on Earth. Most nations have had to defend their territory numerous times against continuing aggression. The War of 1812 with England, during which the U.S. Capitol was burned, was our only major defense, which was coupled with some offensive action (we tried to acquire portions of Canada). Of course, the bloodiest episode in our history was the Civil War, which we fought among ourselves and which led to the abolition of slavery.

Currently, we continue to enjoy an almost unprecedented domestic tranquility. This is the legacy of our victory in wars and conflicts as well as the result of ongoing military innovation, manufacture, and projection of power in the world. While we can argue about what percentage of GNP to expend on the military and what the ultimate goals of diplomacy should be, our ability to address world hunger, AIDS, and provide disaster relief is made possible by the breathing room secured by our military. Without the stable domestic environment it ensures, we would not have the luxury of the time and resources to help others and, more to the point, the freedom to argue that we should be doing more.

From the YANG (left hemisphere) side of the argument:

Conservative (Yang) Declaration: The "bleeding-heart" liberals not only fail to appreciate the necessity of the projection of power and a strong military, they fail to appreciate that the freedom of the individual is of the first importance. This freedom is undermined by the do-gooder socialist programs that address health care, a minimum wage, and welfare. These just create dependency and destroy individual initiative! In like manner, our country needs to remain truly American and deport all illegal immigrants and build the massive and effective fortifications that are required at our borders.

Liberal (Yin) Counter Argument: From the most remote edge of recorded history (4,100–4,040 YBP) we have the first extant set of laws, which were handed down in the ancient Sumerian civilization for the city of Ur and titled the Code of Ur-Nammu. Although this was a highly militaristic culture, King Nammu embraced the concepts of equity and truth, "banished maledictions, violence and strife" and brought "equity to the land."

Beyond the basics of this ancient code, which is considered very advanced in terms of its use of monetary compensation for bodily injury rather than the law of retribution (an eye for an eye); there is provision for widows and orphans, those with the least power to take care of themselves.

King Nammu specifically cast himself as their protector, their benefactor. Even in the worst case of nonobservance, the king clearly felt that this was an important message for the people of his kingdom to hear and for him to profess publicly.

This brings us back once again to the existence of universal moral sensibilities. The formalization of such principles as an integral part of the explanatory narrative was, and is, the essential ingredient in the cohesion of a society, the ability to act in concert, and, ultimately, to act as a superorganism. The willful disregard of neighbors' rights and the acting out on every selfish impulse only leads to anarchy, rape, and murder. Such a society may persist at lower levels of scale and for short periods of time. But the idealized society—which would have internal relations based on the Golden Rule, where everyone is a Good Samaritan—would exhibit the characteristics of self-healing and resilience and could always marshal the highest percentage of the total population to fiercely oppose a common threat. In short, national programs in support of the common good are powerful contributors to national security.

Xenophobia, the fear of those outside one's own race or culture, is clearly a maleness/conservative neurological marker. Since World War II, two nations have stood as polar opposites on the issue of immigration of foreign nationals.

Japan has the lowest immigration rates of any modern nation and now faces severe economic impacts flowing, in part, from a declining and aging population. By contrast, the United States has the highest ratio of immigration, which has contributed to a more balanced and relatively young demographic.

That the United States is expanding diversity in the gene pool while Japan has a stable or declining diversity also has to be considered in light of the fertility rate of the respective countries. In 2012 it was estimated that the United States' rate was 2.05 births per woman and Japan's was 1.39.

Collectively, the effects of immigration in the United States (legal and illegal) are contributing to a more durable demographic profile and a more robust genetic diversity. Again, these are long-term competitive advantages that serve to reinforce national economic viability and security.

To summarize: In the modern era, the liberal, humanistic policies and related expenditures that are attacked by the conservative (Yang), individualistic mindset as wasteful and misguided are often a well-spring of genetic and demographic strength and cohesion that are essential to the readiness and survivability of the nation.

On the other hand, the militarily maintained tranquility in the domestic arena is a precondition for those humanistic policies and the emergence of an equitable, just, and creative culture. The long-term maintenance of national security and domestic tranquility provides the breathing room for the natural tendency of upward evolution—the very objective of the liberal/humanistic mindset.

While these representative declarations and counter arguments are no more than a grain of sand on the beach of world-view differences between the neurological gender left-hemisphere Yang (conservative) and neurological gender right-hemisphere Yin (liberal) sides of our brains, they point to the value of a fully informed and fully expressed debate. In this sense, the two sides constitute an interdependent process in arriving at the correct course of action, or balance of actions. (I say correct in that sometimes there are only "less bad" actions—which, of course, is the correct action). Two interdependent parts of a whole feels intuitively correct.

THE GROUPING EFFECT

By integrating the hemispheres of our brain with only portions of these "DNA sourcebooks" for each human, RNA creates a giant jigsaw puzzle across the population in which there are an unlimited number of unique minds that can fit together for advantageous effect. The insight needed to recognize what those other minds are thinking is called a theory of mind, which we already know as the self-aware mind. It is the three-part combination of diverse minds, the potential competitive advantage to be gained in their collaboration, and the ability to recognize those potentials that collectively create an inherent *grouping effect* across the population. While there are many variants on this phenomenon, one dramatic case in point serves to tell the story.

World War II provides one of the most dramatic examples of how a right-hemisphere Yin (femaleness-dominant) person can be coupled with a left-hemisphere Yang (maleness-dominant) person to achieve in collaboration what neither could have accomplished alone. In the gender realities of 1940, both were anatomical males, but their neurological gender ratios and associated skills were quite different and, as so often happens, complementary.

The Yin exemplar is J. Robert Oppenheimer, who was born into an affluent family in New York City in 1904. He attended the Ethical Culture School and The Fieldston School, both run by the Society for Ethical Culture, whose stated pursuits include "social justice, racial equality, and intellectual freedom." Oppenheimer attended Harvard University, graduated summa cum laude in just three years, and went on to study at the University of Göttingen in Germany, a world-renowned center for theoretical physics. Upon his return to the United States in 1929, he accepted a teaching position at the University of California, Berkeley, and elevated his department to world-class standing.

Oppenheimer exhibited his brilliance in many other facets of his life, including his recognition as an authority on baroque and classical music and his mastery of eight languages. A Democrat, he had a predisposed liberal (femaleness) mindset, and some of his associations early in life would haunt him later during the McCarthy communist witch hunt. Two days after he died, a *New York Times* article on February 20, 1967, reported:

> Beginning in late 1936, Dr. Oppenheimer's life underwent a change of direction that involved him in numerous Communist, trade union and liberal causes to which he devoted time and money and that added to his circle of acquaintances many Communists and liberals, some of whom became intimate friends.

However, at age 41, Robert Oppenheimer was at the height of his career and fame, having served as the crucial intellectual and creative linchpin in the vast undertaking of the Manhattan Project and the creation of the atomic bomb, which led to an almost immediate end to World War II. While in no way prepared or disposed to undertake the staggering logistical, organizational, and political challenge of bringing such a conceptual endeavor into physical reality, he had been paired with a mind that possessed an aggressive, high-energy brilliance of another kind—the hands-on understanding of how to get things done, how to build, how to shape, and how to direct and motivate human potential in the achievement of highly specific and measurable outcomes; General Leslie R. Groves Jr., our Yang exemplar.

Groves was born in Albany, New York, in 1896. He spent his childhood in various parts the United States as his father, a career army officer, was assigned and reassigned to various forts and posts. A graduate of West Point, he finished fourth in his class and was commissioned as an officer in the Corps of Engineers in 1918.

Characterized as a person of imposing personality and iron will, Groves rose to the highest levels of responsibilities for U.S. Army construction in the mobilization period leading up to World War II. A Republican, he had a predisposed conservative (maleness) mindset and was supervising the construction of the Pentagon when he learned that he was to be in charge of the Manhattan Project. The sheer scale of this undertaking, which ultimately required an expenditure of over 2 billion dollars (30 billion in 2014 dollars), and the employment of 125,000 civilians and scientists over a three-year period, had no precedent in the history of the nation.

Lauded for his judgment and his ability to make critical decisions quickly, Groves had a grasp on the enterprise that included the selection of the key production sites at Hanford, Oak Ridge, and Los Alamos; the choice of the key personnel, including military personnel, civilians, and scientists (he even chose Oppenheimer as director); oversight of the test detonation (code name Trinity, in New Mexico); and the operational command of the unit he established to drop the atomic bomb.

The national security state that now exists, for good and bad, was founded in this endeavor and under his leadership. This new reality was characterized by intelligence and counterintelligence, compartmentalization as an organizational concept, and the bridging of efforts among science, industry, government, and the military. For his entire life, Groves was an unapologetic proponent and defender of all things American.

Oppenheimer and Groves, so completely opposite and yet so constructively interdependent in support of their joint mission, stand out among all others as the collaborative essence of the Manhattan Project. It is deeply revealing to note the classical maleness and femaleness remembrances of these men in later years. Richard S. Norris, who wrote *Racing for the Bomb*, observed, "Groves' personal feelings about the bomb were not complicated: He never had any moral doubts, at the time or afterward, about using it." [7]

Oppenheimer's reaction to the detonation was conveyed in his obituary in the *New York Times* February 19, 1967:

And as the black, then gray, atomic cloud pushed higher above Point Zero, another line—"I am become Death, the shatterer of worlds"—

came to him from the [Bhagavad-Gita, the Hindu sacred epic]. Two years later, he was still beset by the moral consequences of the bomb, which, he told fellow physicists had "dramatized so mercilessly the inhumanity and evil of modern war."

In the pairing and accomplishments of these two great minds we see the potentials of assembling the collaborative group and can readily understand the attraction: the rewards to be gained in the reassembly of the "complete mind." Self-awareness allows us to see the advantages of the right mix of those minds, and this skill underpins the rise to the complexity and scope of human endeavor in the modern world.

Even in the relatively small confines of my architectural firm (roughly a dozen persons), I am well aware of those whose artistic bent will assure pleasing outcomes and others who, with their mastery of detail and mathematical accuracy, will excel in the operational aspects of the firm or in the critical review of shop drawings and technical submissions. Other skills, such as marketing and communications, require the rarest of skills in an architectural and engineering endeavor—the ability to write!

The assemblage of a collaborative whole becomes more complex at the scale of larger architectural firms, where one frequently finds hundreds of employees and a triad of principals: one as the "rainmaker" or front person, one as the internal design/artistic talent, and yet another as financial, operational, and project execution specialist. All complex endeavors beyond the scale of the individual depend for their success on the ability to assemble the orchestra of minds capable of producing the music desired.

If the binding of a superorganism in the insect world is chemical and hormonal, the binding effect among humans is common belief and the inherent grouping effect of self-aware minds assembling specialized skills in their pursuit of the complete mind. The self-aware mind has, in this manner, driven the three transitions of scale from the Middle Paleolithic band to the superorganism at nation scale in the modern era.

THE LOYAL OPPOSITION EFFECT

The superorganisms of bee and ant colonies achieve completely robot-like compliance in the pursuit of the common good through chemical and genetic control of behavior. But in a human superorganism how does one address the selfish interests and corruptibility of the self-aware mind, particularly in positions of power and leadership? This is the greatest weakness of absolute power and elite rule, which serves to undercut their economy, and the willingness of third parties to invest as well as undercutting individual motivation and innovation. Most importantly, corruption undermines the commitment and loyalty of the military and the citizenry.

The beauty of an adversarial two-party system within a democracy (particularly when divided along the mind's inborn differences of liberal femaleness and conservative maleness) is that the so called "loyal opposition," the party out of power, serves as a competitive watchdog over the party in power, always looking for any weakness or misstep. The effect of a free press, selective leaks to the media, whistle-blowers, and insiders who expose potential threats to the integrity of the governmental process constitute a complimentary and powerful anticorruption force. While it cannot achieve total elimination of corruption, this transparent and openly adversarial context keeps the incidence of corruption low and under constant risk of exposure. Viewed from afar, this democratic process appears to be anything but orderly and efficient, but it inherently acts to counter the greatest of all flaws, the inherent corruption-prone weakness of elite rule.

THE PARALLEL EFFECT

The same characteristics we see as successful at the various levels of government are realized in many other parallel forms, both corporate and private, but none so rigorously as in the discipline of science. Our nation's scientific community, beneficiaries of the intellectual and social turmoil of the Enlightenment, operates in a state of continuous debate and scrutiny. Every apparent finish line is subjected immediately to an assault in peer-reviewed publications, conferences, and challenges in the public square. Scientists and researchers typically characterize their theories and proposals as arguments, which is to say that everything is part of an open-ended debate, a provisional truth, and therefore able to evolve.

Democracy and science, as conceived in the eighteenth century and implemented broadly in the nineteenth, are evolutionary models that exhibit the characteristics of evolving organisms, and they have come to define the world of the third millennium.

MIRROR NEURONS, MEDIA AND THE MODERN ERA

Humans require a long period of learning, from birth to maturity. Our amount of time spent in formal education, as opposed to learning informally every day, has continued to increase in the modern era. It takes the form of spoken and written language, sensual and cognitive inputs, and imitation of the sounds and actions of others. While their exact role is not fully understood, it is the so-called mirror neurons of our brains that create simultaneous understanding of those incoming concepts and skills, a dynamic often oversimplified as "monkey see, monkey do." However, this parallel-universe effect, which we examined earlier, goes much further than the acquisition of information for an exam or skill for a piano recital.

There is a deep experiential phenomenon by which these neurons fire in our brains in exactly the same locations and sequence as they fire in the people we observe actually carrying out a given action or undergoing an emotional experience. This phenomenon comes into play when we read a book, watch a movie, or observe the distress of others and suddenly find ourselves beginning to tear up or becoming angry at a loss that is not ours but which we have experienced emotionally, not just as an observer.

In the Middle Paleolithic Era, such experiences occurred at the pace of daily life and were limited to the immediate events and surroundings of the observer, but we are free to choose our experiences from an endless selection of media and content delivered when and where we want it via movies, television, computer, tablet, smartphone, or immersive theatrical event. For this reason, the analysis of Internet traffic, Nielsen ratings, and the queries of search engines directly reveal human wishes, wants, needs, and values. These can be reverse-engineered to better understand the human predilections and inclinations that generate those preferences.

If we look at global audiences, the three most-watched TV events of all time are Neil Armstrong's walk on the moon in 1969 (#3), topped by the 2006 World Cup Soccer Final (#2), and the 2008 Summer Olympics (#1). The walk on the moon occurred at midnight local time in Europe and was not broadcast in the Eastern Bloc countries; also some forty years ago, the distribution of television as was not as widespread as today. Therefore, we might fairly conclude that, all things being equal, humans breaking the bonds of Earth to land on the moon would have been the most-watched, most-shared moment in the history of humankind.

However, it is notable that the other two most popular TV shows were sporting events, head-to-head competitions to determine who is the fastest, the strongest, and, by extension, what is the greatest country in the world. The Olympics hosting of the world's most popular competitive athletic events (in which each nation sends the best onto the field under its flag) determines, every four years, which will prevail as the acknowledged champion. In short, it is the civilized version of combat, the most visible demonstration of a nation's courage, and forthright victory that quenches a most demanding human thirst.

Professional football in the United States is the runaway favorite sport, the most-watched event by far, with Super Bowls totaling twenty-two of the top fifty most widely viewed television shows in history! The game is a contest of aggressive, high-impact attacking and defending sequences carried out by teams of eleven male players (approximately the size of a hunter-gatherer band or raiding party). Parallels to the military are everywhere: opposing uniforms and colors, raiding parties moving up and down a field in displays of brute strength, cunning and deception to invade the home territory of the enemy. High levels of violent impact create the same pattern of broken bones and hairline fractures seen in the skeletal remains of *Homo neanderthalensis*. Just as in the Middle Paleolithic, they raid, defend, raid, defend, etc.

First, there are the linemen who, upon the snap of the ball, slam into each other in a battle of pure physical force to control the position of the opposing linemen. These players are physically gigantic, often weighing more than 350 pounds, and massively strong. They are the reincarnation of evolution's most powerful and long-term success in body strength—*Homo neanderthalensis* reborn. Next come the offensive backs and ends (receivers), who, in collaboration with the leader of the attacking force (the quarterback) carry out maneuvers in a combination of deception, speed, and strength. Each of these maneuvers can be simply diagrammed as a series of X's and 0's on a blackboard, with each man having an explicit role in the attempt to penetrate the enemy's territory and to score by raiding his home (the end zone).

The "killer" advantage of *Homo sapiens* over *Homo neanderthalensis* was not only the invention of projectile weapons but the biological advantage of their elongated, slender proportions and the rotational capacity of their extended spinal cord, which, together with long arms, propelled stone-tipped spears great distances downfield to inflict fatal wounds on the enemy. The reincarnation of this body-type and skill is embodied in the NFL quarterback, who throws the football; the long pass down the field; or, in desperation, the Hail Mary pass to gain victory. In this weekly ritual we reenact and celebrate our violently competitive emergence.

Our ability to participate in such an event as a deeply shared emotional experience both during and after the game is confirmed in the unfortunate spillover effect of out-of-control partying, riots, fights, and even shootings that can follow the most hotly contested sporting events, whether professional, collegiate, or high school. Nevertheless, this channels our need to compete physically, under the colors of our tribe, in a recreational rather than deadly pursuit of victory.

The popular desire for competition with clear winners and losers is reflected across the home entertainment spectrum, listed here in no particular order: *Top Chef, Dancing with the Stars, Chopped, American Ninja Warriors, The Biggest Loser, The Voice, The Bachelor, Iron Chef, Sweet Genius,* and *Cupcake Wars.* These highly rated winner/loser programs, as a way of stepping up the intense audience engagement, are starting to use Internet voting to select winners by the audience!

Creating a deeply felt experience worldwide without the necessity of a physical presence has created the phenomenon of the virtual cultural invasion.

No longer is it necessary to steal women and occupy land to "Neolithize" the hunter-gatherers. This invasion occurs through Hollywood movies, the Internet, Twitter, television reruns of *Seinfeld* and *The Honeymooners*, Apple, Microsoft and IBM, CNN and McDonald's, and a thousand other avenues (in these American examples) that flood the global bandwidth—economic competition with a deep cultural side effect.

In this way, media becomes the vehicle of language, style, values, and culture that pass over, under, and around a nation's defense system, arriving fully formed in the eyes and minds of the audience. A very popular viewpoint is that we are at the end of the "American Century," which may very well be true in terms of GDP, but who will suggest that the new global language in this century will be Chinese, or Russian, or Japanese? There is a large American cultural signature and momentum interwoven in global culture (for better and for worse) that will carry on for some time, essentially indifferent to the economic fate of the United States.

The reality is that we are deeply into the Third Transition, that is to say, the death of elite control of the explanatory narrative. The remaining regressive (typically elderly and male) commanders of the Second Transition model do not want to surrender control. The younger generations in China, Russia, Egypt, Iran, and Afghanistan, those most attuned to the social and cultural river of information flowing through the Internet, will ultimately lead in the process of change. Russia's and China's need for the Internet will, in spite of censorship, continue to open a window to the free-market democracies. As mentioned earlier, the Organization for Economic Co-operation and Development (OECD) substantiates that their organization of 34 democracies, representing only 18 percent of the world's population, produces 63 percent of world GDP and 75 percent of world trade.

ISIL is the exception that proves the rule. Benefitting from the rampant corruption and elite pro-Shiite bias within Iraq's government and military leadership, they have successfully exploited the Internet and social media for amplification of their message. ISIL is inflamed by its beliefs and righteous indignation, whereas Iraq's soldiers have no belief or trust binding them to their leadership and nation. As previously observed, this ISIL formula can work in the short term, but in the long run, theirs will be the fate of all such brutal, male-dominated and violently repressive regimes. Pussy Riot, the feminist punk-rock band in Russia, is a more prophetic indicator of the direction we are headed internationally on the matters of personal freedom and the power of modern media.

In short, rising transparency and accountability, coupled with humankind's continuing biological transition to the creative tendencies of right-hemisphere thinking, are consistent with the continuing decrease in violence and the rise of democracy in resources and influence across the globe.

OVERSHOOT

A final characteristic that has all of the markings of a deeply rooted biological behavior is the tendency to fully exploit every opportunity, even to the point of collapse. Jared Diamond explores the phenomenon and its tragic consequences at the larger scale of human communities in *Collapse: How Societies Choose to Fail or Succeed.* Overpopulation beyond the carrying capacity of the surrounding environment was a central theme of the failures Diamond cites, but we see this behavior expressed in many others ways in our modern world. Some examples are the pattern of real-estate overbuilding and market collapse, stock market swings from record highs to lows, even patterns of overeating and the resultant overweight Western populace.

No matter how many times we repeat this boom-and-bust pattern, or how sophisticated we imagine our subsequent analysis to be, there seems to be no learning from past experience. The evolutionary source of this behavior is seen in the classic expansion of species to the point of resource depletion, or the boundary of a competing species, or the failure of an adaptive advantage to work within a new environment. In other words, there is a gas pedal but no self-regulating brake in ecological systems. Life expands until there is the feedback of hitting a wall.

As long as our presence was at animal scale or hunter-gatherer or even agrarian scale, overshoots were the mistakes of the local tribe or society, which had to pay the price. Now, with the advent of our discovery and unrestrained consumption of Earth's multibillion-year carbon bank (fossil fuels), we are forcing an energy imbalance at planetary scale, driving a slow-motion (low-inertia) global change that has a point of no return long before we hit any tangible wall. This carbon-consumption overshoot will bring a tragic reckoning to all of humankind in the generation of our children, grandchildren, and beyond unless we act now— quickly and decisively—on the basis of the knowledge we have in hand.

The heavy consequences, which are no mystery, having occurred over and over again in the geologic/oceanographic history of Earth during well-documented warming periods, *have never happened during humankind's brief history on the planet.* These consequences would include the loss of the world's coastal cities, instability in the face of massive population dislocations, cross-border incursions, and the undermining of food and water supplies.

The human tragedy—suffering and death at large scale—that most climate scientists see as currently growing from probability to certainty may already be unavoidable, according to several credible studies, first among them "Assessing Dangerous Climate Change" by James Hansen, Pushker Kharecha, et al.[8]

This is the ultimate test of the self-aware mind and the innovative potentials of humankind because, as of this writing, we have come to a final threshold in our need to act. The UN Conference of the Parties on Climate Change in Paris in December 2015 will be targeting an international agreement by which all nations will take the actions necessary to hold global temperature rise to 2 degrees centigrade (3.6 degrees Fahrenheit). This action is viewed by many scientists as too little too late while contrarians on the left-hemisphere (maleness) side of the argument insist on waiting until we hit a tangible and irrefutable wall before they will believe or act, long past any chance to avoid disastrous consequences and stabilize the climate and Earth's thermal balance.

The massively warming oceans and ice sheets are already loaded with heat input and are progressively acquiring the inertia to deliver this new reality to future generations—a reality that, even now, we see unfolding around us. Here, at the height of our evolutionary advance, we must step fully into the promise of our self-aware minds and exercise our free will based on acquired knowledge. In short, we must disregard the impulses from an ancestral world that has no relevant messages for this moment.

A CIRCUMSTANTIAL CASE

This book started as a personal opinion in 2002, key concepts were sketched out in 2005, and final development began in 2010. I am not a scientist but a practicing architect with a lifetime of experience in the environment, human behavior, and integrative thought. This was the basis of my confidence, my belief that I had a valuable insight that needed to be fully developed.

Spurred on by my personal and professional insights but recognizing the need to test and ground my assertions over time, I have closely followed the emergence of relevant findings in the scientific world, all the while advancing what I believe to be a compelling circumstantial case for the roles of DNA and RNA in the long-term and short-term resilience of the mindset guiding our species, and the centrality of gender in the creation of individual personalities, talents, and worldviews. I never expected there to be a "CSI moment" with the discovery of forensic evidence and the attendant analysis of DNA to support this point of view!

In the summer of 2014, just prior to publication, I completed my final review of the scientific literature and discovered a paper at the molecular level of the DNA of *Homo neanderthalensis* published by Sriram Sankararaman and David Reich and their team. To date, this has to be the greatest moment of affirmation and discovery in my professional life. I share the details of this finding in Chapter 6: The Revelation.

CHAPTER SIX

The Revelation

I HAVE ALWAYS BEEN CONFIDENT THAT THE ASSERTIONS PUT FORTH IN this book would, in time, find scientific validation. Over the years required to complete the manuscript, a number of my initial premises were confirmed through the work of others whose publications I folded into the book with footnotes as I went along. However, there remained a key assertion that I knew was unlikely ever to be validated in spite of its profound significance, because it was prehistoric in origin and subtle in neurological expression.

The assertion that natural selection, through DNA, creates two separate repositories of male and female life experience in the form of the two hemispheres of the brain is an example of a premise that can be scientifically proven, one way or the other, in the near future. However, it seemed beyond any possibility that the emergence of the self-aware mind, 100,000 years ago, utilizing RNA pathways to expand the interconnectivity of the gendered hemispheres of the brain of *Homo sapiens* versus those of *Homo neanderthalensis*, would ever be scientifically validated.

My focus on the significance of the final confrontation of *Homo sapiens* and *Homo neanderthalensis* grew from the strong suspicion that such a difference—a neurological nuance of infinitesimal scale—had allowed our ancestors to cross over to self-awareness and prevail over a physically superior adversary. My strategy was to build a supporting case based on the behaviors of these Middle Paleolithic opponents and to close the argument using the mapping of the brain as advanced in the contemporary neurological sciences. The intent was to prove, circumstantially and beyond reasonable doubt, that this was the case. I never imagined that physical evidence of such a difference would be uncovered!

"The genomic landscape of Neanderthal ancestry in present-day humans" by

Sriram Sankararaman et al.[1] was published in *Nature* on March 14, 2014 (online on January 29, 2014) By examining the genomes of over a thousand present-day humans and finding genomic regions that were high or low in the small amount of Neanderthal genetic material we still possess, scientists were able to identify various areas that were highly resistant to the interbreeding of *Homo sapiens* and *Homo neanderthalensis*. In other words, they were able to understand specific features in modern humans that were absent or very different in *Homo neanderthalensis*.

For instance, the consistent indication on the modern day X chromosome of very low *Homo neanderthalensis* ancestry is convincing evidence of male hybrid sterility in the offspring of any such interbreeding in the Middle Paleolithic. A similar effect is observed when a horse and a donkey breed across species to create a mule. *The male mule is born sterile and cannot reproduce.* Such a finding is consistent with the degree of divergence we have asserted for the 458,000-year separation of these Middle Paleolithic opponents from their common ancestor.

However, the most stunning revelation was near the end of the paper:

> Hybrid sterility is not the only factor responsible for selection against Neanderthal material as **Neanderthal ancestry is also depleted in conserved pathways such as RNA processing**. (Author's emphasis)

I read those words but couldn't fully process them; and then I read them again in disbelief. *Beyond male hybrid sterility, only RNA was specifically identified as an overall factor in the incompatibility of the two species!* I was taken aback that something so significant was mentioned incidentally in a paper on other matters.

Sankararaman and his colleagues included a parenthetical note referring readers to "Section 6 of Supplementary Information." I immediately contacted *Nature* to get what turned out to be over 100 pages of supporting documentation and there, on pages 58–60, among the sixty-five components and functionality with low "Neanderthal ancestry" values (Tables SI 6.1 and SI 6.2), were the following four ncRNA dimensions that are key to brain development and behavior:

1. RNA binding: Molecular function (code: GO: 0003723)
   Baseline ncRNA requirement
   (Including tRNA, miRNA, siRNA, piRNA and snRNA)
2. RNA transport (code: KEGG 3013)
   Baseline ncRNA requirement
   (Including tRNA, rRNA, and snRNA)
3. Cysteine and Methionine metabolism (code: KEGG 270)
   (Foundational to ncRNA tools/process for "methalation," "expression" of genes in brain development and behavior)

4. Hedgehog signaling pathway (code: KEGG 4340)
*(Key director of embryo/fetal development from stem
cells to tissues and organs, e.g., the brain)*

The fact that RNA molecules and functionality in these categories were listed in the low "Neanderthal ancestry" tables, indicates that they were either not present or that there were significant biological differences between *Homo neanderthalensis* and *Homo sapiens*. The second-lowest of sixteen tissues identified in the "Neanderthal Ancestry" tables were the tissues of the adrenal gland which are involved in the *development of brain cells that establish sexual orientation and gender identity*. (See the discussion in Chapter 4 of the endocrinology study by Garcia-Falgueras and Swaab.) The adrenal gland produces male and female sex hormones (testosterone, estrogen, and progesterone) central to the development of brain and gender. Unlike other organs of the endocrine system that communicate through the bloodstream, the adrenal gland has a nerve connecting it directly to the brain.

In short, the overall role of RNA in interconnecting the hemispheres of the brain, as well as the adrenal gland and the regulatory role of the endocrine system are, to the highest degree of probability, unique to *Homo sapiens*. Self-awareness and identity, so closely intertwined with the enriching effects of the neurological gender ratio would therefore be completely absent or significantly diminished in *Homo neanderthalensis*.

The location of these compelling differentials at the "center of gravity" of self-awareness and identity constituted, to this author at least, a personal revelation and the first scientific evidence supporting the *Two Minds* thesis. This amazing turn of events was to serve as the dramatic conclusion for the book.

NOT THE END

On February 28, 2015, with the addition of ten summary conclusions, I finished the book and sent out six review copies in hopes of winning support for my effort and a couple of good quotes for the book's dust jacket. My editor, Janet Adams Strong, was concerned that I had included two scientists as reviewers and advised me to brace for the response because, by definition, I was clearly the uninitiated encroaching on an infinitely complex field without the training and language to speak in their voice.

She was right. Dr. Peter H. Raven, chairman emeritus of Missouri Botanical Garden, and Dr. Carl N. McDaniel, a developmental biologist and professor emeritus of biology at Rensselaer Polytechnic Institute, and the author of *At the Mercy of Nature* (2014) had a resounding "Stop the Press" response, which I heeded.

Drs. Raven and McDaniel each insisted that I could not posit DNA and RNA as the opposite domains of the human genome as I had done in the review version; rather, they should be designated "protein coding DNA" and "non-protein coding DNA." Therefore, I have gone back through the book to address this distinction and other insightful suggestions. However, as the interpreter of DNA in matters of the mind, I have retained RNA (ncRNA), whose control over matters of perception and behavior has recently been stated as "abundantly clear" in the article "Noncoding RNAs and neurobehavioral mechanisms in psychiatric disease," for *Molecular Psychiatry* (March 31, 2015) by J. Kocerha et al.[2]

Dr. McDaniel had more fish to fry, and concluded a nearly line-by-line full-throated critique. While I have not taken him up on his recommendation for an additional year of study in the field of Developmental Biology, in the process of pursuing his challenges, I stumbled on an entirely new avenue of validation.

Dr. McDaniel's most specific challenge to my conclusions centered on Sankararaman's paper in *Nature* and the need for greater detail in interpreting the significance of the genetic remnants of *Homo neanderthalensis* in modern-day humans. Therefore, my first step was to go back to the 101 pages of supplemental information supporting the paper to report the depth and relevance of the findings more fully. These important elaborations have been incorporated in this chapter and, in seeking those answers, a casual question led to the most important revelation of the book.

MIRIAM'S QUESTION

Since Sankararaman's paper listed sixty-five factors with low "Neanderthal ancestry" (most of which were unknown to me), I undertook to have a scientist in the field of genomic biology independently evaluate their relevance to *A Convergence of Two Minds*. The first expert who came to mind was a recent client, Eva B. Cramer, Ph.D., the director of the Biotechnology Incubator at SUNY Downstate Medical Center in Brooklyn, New York.

Eva very quickly concluded that this was a question of genetics and molecular biology and introduced me to Miriam H. Feuerman, Ph. D., in the department of cell biology. In our first informal conversation on the nature of the book and my initial description of DNA inheritance and RNA as "interpreter" of DNA—creating personality or *personhood* without changing the underlying DNA—she asked, "Are you familiar with epigenetics?" My initial response was yes, although I had a vague recollection that epigenetics referred to the inheritance of DNA modifications that had come about as a result of events occurring in preceding generations, such as famine or child abuse. Therefore, I doubted that there was a connection between epigenetics and *A Convergence of Two Minds*, which is about identity which is unique and cannot be inherited.

Dr. Feuerman then forwarded for my further consideration (read: education) the PowerPoint slides used in her genetics class for pre-med students, and I separately dug into an introduction to epigenetic biology.[3]

For all of my confidence in the assertions of the Two Minds Theory, I knew that most of the evidence is circumstantial and that without finding more specific biological pathways for the phenomena within the scientific record, it would remain, for most, no more than an interesting speculation. The task of finding those pathways seemed insurmountable when I considered the biological gymnastics that would be required to create the outcomes I was proposing. Among the many unresolved questions I had, these were the most puzzling:

1. Where does this purely male and purely female legacy of human experience exist, given that we are a combination of two sets of male and female DNA—one set from our mother and one set from our father? How could only one set of each gender's ancestral line be retrieved from these four to form the hemispheres of our mind?

2. Even if we achieve our biological uniqueness, our identity, by trimming such a massive male and female ancestry down to the human-sized subset of an individual, how is that original legacy maintained so that it can be passed on?

3. If we each possess a unique biological version of our identity, where did the versions of our parents go and when and how is such an identity created for our children?

Epigenetics, contrary to my expectations, proved to be the most relevant of all possible disciplines. As currently conceived, it is primarily focused on heritable effects and is not relevant to "two minds." Nonetheless, epigenetic science has advanced a highly relevant grasp of the mind's biological toolkit. The focus of epigenetics is the interpretation or "expression" of DNA (or a strand of DNA called a gene) that occurs without modification to the underlying structure of the DNA. The primary tools for this "expression" are small epigenetic markers and tags that, for example, can be applied (methylation), removed (demethylation), and applied again (remethylation) by the legions of RNA, all without changing the DNA.

Much as a rider puts bridle, bit, and reins on a horse to direct it when and where to go, RNA can fit out DNA and/or genes for a wide range of functional purposes as intended without changing the "horse." I quickly recognized the explicit capability of a robust epigenetic process to orchestrate the immensely complex interconnections and influences between the hemispheres of the brain.

With this powerful tool available at the molecular level, RNA can, through epigenetic processes, "interpret" the *DNA brain* to synthesize a neurological gender ratio, NGR, for one generation and then (in the process of the same birth) remove that interpretation in order to "interpret" a new NGR for the reproductive cells of the fetus for the next generation. The underlying *DNA brain* remains unchanged and passes forward through the generations with only the slow modifications of natural selection (see Figures 4.5 & 4.6). Each epigenetic interpretation of the hemispheres is a new "applied" NGR, which, combined with our free will, constitutes our identity, a unique personhood, for a lifetime.

As I read more deeply into the emerging science and processes of epigenetics, a more direct connection to my theory became evident. While epigenetics applies to the entire human body (the full complement of approximately 22,000 genes within the human genome), there is a highly specialized epigenetic process called *genomic imprinting* that applies to only a few hundred of those genes. These are the genes primarily involved with the growth of the embryo/fetus from conception through the neonatal period twenty-eight days after birth, as well as brain development, the neurological system, and behavior, in other words, *everything at the center of our attention.*

Genomic imprinting in the formation of your mind begins with the creation of your parents' sperm and egg at their birth that, many years later, will be joined to form you. This is an event that poses the central unanswered question in the field of epigenetic biology. In the creation of the single-cell sperm and the single-cell egg in your parents, only one half of each parent's DNA pair remains functional, *the other half is "silenced."* In the sperm, the male half of your father's DNA remains functional, and in the egg, the female half of your mother's DNA remains functional. Why?

Three explanatory theories have been put forward over the last twenty-four years, summarized in 2014 by M. M. Patten et al in *Nature* in "The evolution of genomic imprinting: theories, predictions and empirical tests." [4] All three theories are cast in the one-dimensional framework of natural selection. To the contrary, this is the basis of neurological diversity where all ancestral male and female experience has been separately aligned in egg and sperm, passing directly into the two separate strands of DNA joined at your conception—"two minds."

In this manner, genomic imprinting connects a continuous (and separately maintained) genetic line of male ancestral experience and a continuous (and separately maintained) genetic line of female ancestral experience that are ultimately embodied as the two hemispheres of the human brain—the physical "mirror" of this union.

At this point, there should be a bolt of lightning and a roll of thunder, because in this simple epigenetic move, millions of years of viable male and female survival behavior has been linked and made accessible. This is the foundational requirement for the two phenomena, genetic and epigenetic, of *Two Minds*.

RECONCILING EPIGENETICS AND "TWO MINDS"

In Figure 6.1, we juxtapose the formation of the human mindset (a process spanning four generations and three births) with the milestones of "two minds." This is consistent with our two basic premises:

> P_1. Humans possess separately conserved male and female sourcebooks of ancestral knowledge in the physical form of the two "gendered" (maleness and femaleness) hemispheres of the brain.

> P_2. The identity or *personhood* of each individual is achieved in the integration of these gendered hemispheres. This neurological gender ratio (NGR) exists in only one person and cannot be inherited. (The author's family is used for purposes of illustration in Figure 6.1).

Step 1: Ancestral DNA
The first step in the creation of your mindset occurs during the birth of your parents. During the formation of the reproductive organs of each parent, there is also the selection of the genetic information to be passed on by the sperm on your father's (paternal) side, and by the egg on your mother's (maternal) side. The sperm consists of a single cell that contains only the male half of your father's imprinted DNA (which he had received from his father) and the egg consists of a single cell which contains only the female half of your mother's imprinted DNA (which she received from her mother). Genomic imprinting, in effect, removes your paternal grandmother (as happened with all paternal female ancestors) and your maternal grandfather (as had happened with all maternal male ancestors); you inherit *a maleness ancestry of experience and a femaleness ancestry of experience*. The two collective wisdoms of humankind (male and female) passed down to you intact! (P_1)

Steps 2 and 3: Erasure and New Identity Potential
The second step is the erasure of identity (epigenetic markings) from the half-DNA cells of the sperm and egg. This process strips away the identities of the previous

owners (your mother and father) from each half-DNA cell down to its essence: the *primordial germ cell*. Finally, the baseline paternal and maternal epigenetic markings (you) are placed on the half-DNA cells of your parents respective sperm and egg. $(\mathbf{P}_2)$

Steps 4-9: Fertilization and NGR 1

Many years later the sexual union of your parents will start the sequence of your birth. Following fertilization, there is a merging of your mother's and your father's versions of you. This immediately proceeds in a deconstruction (removal and reading) of the epigenetic markers of your father's version of you and a deconstruction (removal and reading) of the epigenetic markers of your mother's version of you. Upon reaching a common level of being "unmarked" DNA, the two versions of you are integrated as one unique neurological gender ratio, NGR 1, through a process of reconstruction (remethylation). All the connections and subtle influences between the two halves of your brain are mapped in this first week and are carried out over the forty weeks of gestation. This is you, for life, unique and, as a whole, not heritable. $(\mathbf{P}_2)$

Steps 10-12: NGR 2 for Your Offspring

The final step in the process of your birth is the formation of your sperm or egg. Your identity, NGR 1, is removed and NGR 2 is applied for the next generation—your future offspring.

It is puzzling that epigenetics is generally thought of as being heritable. Identity cannot be inherited, since it serves as the near-immutable neurological signature for the opposite purpose: diversity. It was therefore a great relief to find the National Institutes of Health's (NIH) recently modified definition of epigenetics in the NIH "Roadmap Epigenomics Project," as follows:

> For purposes of this program epigenetics refers to both heritable change in gene activity and expression (in the progeny of cells or of individuals) and also stable, long-term alterations in the transcriptional potential of a cell that are **not necessarily heritable**. [Author's emphasis]

Within the Two Minds Theory, genomic imprinting serves the intergenerational purpose of protecting and projecting these several hundred genes independently of the larger genetic and epigenetic processes involved in the creation of the body. Without it, our complementary dimensions of resilience—one of convergence (natural selection) and one of divergence (neurological diversity)—could not

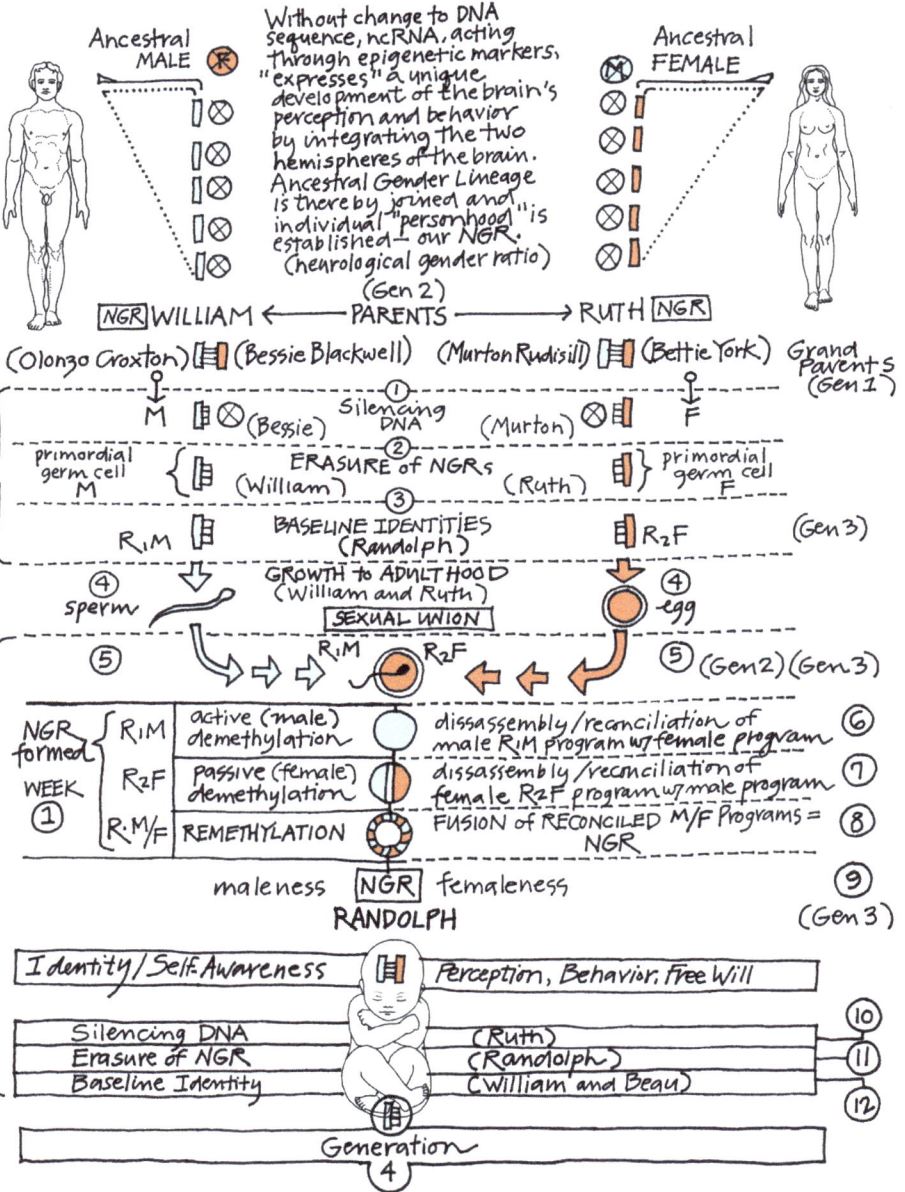

Figure 6.1. Proposed Genetic and Epigenetic Pathways
in the Formation of Our "Two Minds"

coexist as we observe and experience them, our *Two Minds* in near-continuous deliberation with our free will, our way of being in the world: our identity.

The advent of epigenetics and the near-infinite permutations of "two minds," can easily account for the individuality we observe in the 7 billion people on Earth. As Dr. Ian Cowell, writing for the British Society for Cell Biology, states, epigenetic modifications "represent an almost unimaginable amount of information, dwarfing even the human genome project." Dr. Cowell goes on to liken the task of mapping epigenetics to the immense scale and complexity of the Large Hadron Collider (LHC), the massive particle accelerator being used to explore the smallest, subatomic particles of matter.[5]

While it is theoretically possible that another person, exactly like you or me, could be born at some future point, the statistical probability that such a thing could happen even between now and billions of years hence, when our sun explodes is, I would suspect, statistically improbable. We each represent a unique state of self-awareness and identity that, likely, will never happen again.

SIGNIFICANCE AND MEANING

We have asserted a most personal reality: that we are not under the tyranny of our family DNA, nor are we following the strict dictates of our genes. We have asserted that each one of us perceives the world as no other human has, or likely ever will—a uniqueness that stands in contrast to society's view of "normal" because, at the most fundamental biological level, there is no such thing as "normal".

We have proposed that there is an ancestral chorus animating the two sides of our internal conversation which, over time, has to negotiate with the voice of our free will. Eventually, we ignore the worst of this ancestral advice and bend the curve of the slowly evolving DNA brain; in this manner we add our voice to the descending ancestral chorus.

The first moment of our full awareness—the first moment of our humanity—followed the joining of the creative and empathetic dimensions of our being (femaleness) with the protective and provisioning dimensions of our being (maleness). The evolutionary pathways from that beginning have left me with a deep appreciation for the interdependence of these two natures, particularly the maleness features which I previously did not fully appreciate. As I watch coverage of the beheadings by ISIS and of the kidnapping and abuse of schoolgirls by Boko Haram, I am reminded of the necessity of this balance.

While I never expected to be drawn so deeply into the molecular foundations of our awareness and identity; the resulting insights into human resilience—

genetic and epigenetic—have provided a deeper understanding of our humanity and dominance.

Many new perspectives flow from this framework, and there are four aspects that call out for reconsideration:

a. **Gender:** Understanding human gender has been a central objective of this effort. The biological pathways and expressions of gender that we have uncovered confirm that male and female attributes are no less than the source material of perception and behavior.

Due to society's singular focus on gender in the context of sex and reproduction, our cultural heritage of philosophical, artistic, and scientific accomplishment has been thought of as an unrelated dimension—cerebral and academic. Quite the contrary; the conceptualization of music and literature occurs on the "femaleness" side of our brain, while the analysis, rigor and execution, are a matter for the "maleness" side. In this manner, neurological gender is the well-spring of our creative and perceptual abilities and is foundational to the artistic, scientific, technological, and philosophical accomplishments of humankind.

Edward O. Wilson in his recent book, *The Meaning of Human Existence*, describes an intriguing parallel to *Two Minds* in which he identifies an internal turmoil—the "opposing two vectors" that are "hardwired in our emotions and reasoning, and cannot be erased." Wilson identifies this conflict in the mind as being "caused by competing levels of natural selection." One of these forces he attributes to natural selection's rewarding selfish behavior (we call sin) of the individual acting within the group while the opposing force he attributes to noble behavior (we call virtue) of the individual acting group-to-group.

In contrast, the Theory of Two Minds asserts that this internal conflict is a deliberative feature of human resilience that simultaneously accesses the respective maleness and femaleness ancestral wisdoms of humankind. These forces of the mind are seen as opposing yet interdependent and, in many ways, subject to the exercise of free will.

In his 1998 classic, *Consilience: The Unity of Knowledge*, Wilson called the linkage of the Sciences and the Humanities "The greatest enterprise of the mind…" In a sense, this linkage is created in the birth of each human mind. RNA through the epigenetic tools of markers and tags creates a unique reconciliation between the analytical/object-oriented (Sciences) maleness of the left hemisphere and the creative/social (Humanities) femaleness of the right hemisphere. The living of each life holds the potential for the reconciliation of the Sciences and the Humanities because a single mind—as in the exceptional example of Edward O. Wilson—may span both branches of knowledge.

b. **Diversity:** There are geopolitical and strategic implications in the way gender and diversity we embraced in a nation's culture. Viability for a nation in the modern world is reinforced when individual ways of perception are recognized as inborn dimensions of resilience across the population.

   As a model of national consensus and security, a democracy assembles diverse minds in a deliberative congress to recreate the DNA "complete mind." In this manner a nation can approach the highest level of social order, the superorganism. However, a complete version of this national ideal has yet to be achieved due to the influence of special interests and the failure to achieve a fully representative (diverse) participation in governance.

   By recognizing diversity as an inherent asset, a nation will attract the genetic and cultural resources of other nations, as well as investment capital, and a competitive advantage at global scale is created. Embracing diversity and attracting diversity is central to the long-term survival of a nation in the modern world.

c. **Free Will:** Finally, there is the reality that our continued existence is threatened! We have continued the ancestral drive for growth and expansion into the modern era without acknowledging the well-understood consequences. We have reached the limits of our instinctual drive to possess and fill every niche. We have pushed the thermodynamic equilibrium of Earth to the threshold of a global reversal. Our aggressive pattern of pressing on to the point of failure—be it the stock market, real estate, or financial bubbles of recent creation—is now driving us to the precipice of slow-moving and unforgiving planetary forces.

   All previous consequences of our tendency to overshoot were relatively local and, over a period of time, reversible. The approaching events—the inundation of coastal cities, the widespread migration and dislocation of human populations, and the erosion of food resources—are neither local nor reversible in any meaningful time frame for our civilizations.

Rising to this challenge requires that we step out of the conversation with our ancestral voices that has propelled us to this point, and engage in the informed contemporary voice of our free will. By acting on available knowledge of sustainability and the informed pathways of science (metrics and methods), we can maneuver around this self-inflicted global crisis.[6]

d. **RNA World:** Because our central premise is that RNA (epigenetic), is determinant in the creation of the mind, I characterize the Two Minds Theory as an Epigenetic/Genetic theory of human emergence. This brings us back to an earlier theory, an "RNA world," which is the proposition that RNA existed as the first formation of life on Earth and that DNA followed.

As asserted here, RNA is assembling and projecting the unique identities of humankind in each generation—an instant and non-adaptive dimension of resilience across the species. DNA appears to be playing a supportive role in this dynamic.

For all the characterizations of DNA's "selfish" genes and their willful actions in the control of our destiny that have dominated recent thinking in the natural sciences, genes turn out to be (at the level of the mind) more like birthday ponys that are being epigenetically bridled and saddled by the tags and markers of RNA for their creative pursuits in forming the human mind—a revolutionary reconceptualization of gender and human origins, should it prove to be correct.

TEN INTERLOCKING CONCLUSIONS

1. **The Hemispheres of the Brain Constitute Two Minds:** From the beginning of the genus *Homo*, over seven million years ago, the survival pathways of our male and female ancestors (two very different ways to stay alive) continued to be passed down, through natural selection, as separate genetic legacies, or sourcebooks—the hemispheres of the brain.

During this prehuman phase, the anatomical gender (male or female) was also the dominant hemisphere of the brain. As with all members of the genus *Homo* at that time, a powerful dominance of males, and therefore the "maleness" hemisphere, was the biological reality reflecting the survival imperatives.

2. **The Self-Aware Mind**, which marks our emergence as humans, began approximately 100,000 YBP which was 100,000 years after our species emerged. This birth of humankind was the result of a biological process which was coupled with the competitive advantage *Homo sapiens* gained through the innovation of superior weaponry—most logically, projectile weapons. Exploiting this advantage to eliminate less fully aware competitors and thereby expanding their own range, population, and resources, *Homo sapiens* were the first of the genus *Homo* to achieve the fully self-aware mind.

    This phenomenon was a biological feedback loop, initiated, by the evolutionary imperative of competition and concentrated through natural selection. Notably, it originated as a greater expression of femaleness (creativity), reflecting an increasingly interconnected left hemisphere (maleness) and right hemisphere (femaleness) in males and females.

3. This new level of **Interconnectivity of the Hemispheres of the Brain** was the determinant biological factor in the emergence of the self-aware mind in *Homo sapiens*, and, was the source of their competitive advantage employed to prevail as sole survivor of the genus *Homo*. This evolutionary breakthrough was a neurological function of RNA (epigenetics) acting in counterpoint to DNA (genetics) in the formation of our *Two Minds*.

4. **A Unique Way of Being in the World** for each individual is cast by RNA (epigenetics) in the selective connection of the two DNA gendered hemispheres. The resulting mindsets, a vast range of maleness-to-femaleness ratios, constitute a distributed (and evolutionarily advantageous) pattern of variance in the population that is achieved generation after generation. This integrative RNA epigenetic function creates in each individual a unique neurological gender ratio, (NGR), through a process of neurological gender synthesis (NGS).

5. **Resilience of the Human Species** is a function of both DNA/ natural selection (genetic) and RNA/neurological diversity

(epigenetic). DNA imparts resilience at the scale of the individual by assembling the "sourcebooks," the total maleness and femaleness potentials from which a mind is then composed. The epigenetic process is the source of resilience at the scale of the species due to the diverse range of NGRs which create wide-ranging perceptions, and, thereby different responses to threat and opportunity.

6. **DNA (Genetic) and RNA (Epigenetic) Are Independent But Complementary in Function**. DNA is informed by past events (natural selection) and has the effect over thousands of generations of converging physical attributes toward an ideal fit to the existing environment. By contrast, RNA has the effect of anticipating future events by creating a resilient population of diverse worldviews, sexual preferences, talents, skills, and predilections (neurological diversity) upon birth, so that in every generation there will always be those most fit for an unknowable future.

    Note: The RNA diversification effect is not subject to the DNA/natural selection effect, and is, therefore nonadaptive.

7. **The Specialization of the NGR Mindset Creates a "Grouping Effect."** Each NGR (each person) possesses only a portion of DNA's "complete mind." The total range of minds is like a jig-saw puzzle of many pieces drawn from a larger picture. Through what is known as a Theory of Mind, humans are able to anticipate what others are thinking and, by extension, recognize potentially valuable features in the minds of others. Perceiving potential competitive advantage, humans act to form alliances and collaborations.

8. **A Self-Righting Effect Is Created by DNA/RNA.** In the aftermath of a cataclysmic event such as a meteor strike, massive volcanic eruptions, or severe climate change with a multiyear nuclear winter, all femaleness-dominant males and females are at high risk for early death. Such a post-apocalyptic and hypercompetitive environment will be survived predominantly or even exclusively by the most extreme maleness-dominant males and females.

    RNA/epigenetic pathways in each subsequent generation will constantly reintroduce the diversity of maleness/femaleness mindsets even if the surrounding environment remains hostile and none of these humans who are femaleness-dominant survive to reproduce.

And this is the key: femaleness-dominant humans within this deadly environment are not "fit" in the Darwinian sense of the word and, if this environment were to persist over thousands of generations under natural selection alone, this femaleness attribute of humankind would eventually be eliminated. However, independent standing in the RNA/epigenetic pathways assures the continuation of the range of NGRs (maleness/femaleness) in a neurologically diverse and resilient descendancy of humankind.

When an environment of domestic security, stable climate, and green Earth is reestablished, males and females of that generation, possessing the femaleness qualities of trust, creativity, and opportunity bias, will create value and be valued and will rise to the fore. Capitalizing on the expansion and prosperity potentials of such a moment is essential for competition and, therefore, survival and returns humankind to its overall pattern of upward evolution.

Note: In the face of warfare, the prepositioned maleness-dominant minds are there to rise to defend the society, assuming their worldview has not been fatally marginalized. The lesson in all this is the vital interdependence of two "opposite" minds.

9. **God: The First Indicator of Human Intelligence.** God, gods, and religion emerged in parallel with self-awareness in humans (which was the ability to see oneself in the context of the larger universe and recognize the ultimate mortality of humans). These were the first of the necessary explanatory narratives (removing fear: a maleness imperative) and the associated formalization of moral sensibilities (empathy: a femaleness imperative) that provided the binding force of common belief necessary to move early humans from the scale of the band to the level of civilization and ultimately to the threshold of the superorganism.

10. **Democracy: The DNA Complete Mind and the Superorganism.** By creating, and regularly recreating, a broadly representative national congress of minds in the open competition and advocacy of a two-party system (a replication of the opposing poles of the

Figure 6.2. *Homo sapiens* Cave Art: Facing a charging Deer Herd (23,000) YBP

*Note: Elongated body types and mastery of the long bow in the hunt,
the signature feature of the Homo sapiens hunter-gatherer.*

brain), democracy functions in an evolutionary manner.

A free press and the competition of parties in and out of power creates an unruly but anti-corruptive context, a defining advantage over elite rule and its inherent tendency toward corruption.

In this manner, a democracy can operate as an evolutionary model, upwardly evolving and constantly aware and responsive to surrounding conditions, approaching the realization of a superorganism but never reaching the insect world's robotic compliance (achievable only in the absence of free will).

In closing, I propose the following amendments to our current definitions of Evolution, Gender, and Free Will in the matter of our minds:

EVOLUTION

DNA/Natural Selection, a genetic phenomenon at the center of human physical advancement which captures the two streams of ancestral fitness as the male-influenced and female-influenced hemispheres of our brain (heritable). The positioning of these hemispheres within the brain is maleness-left hemisphere, and femaleness-right hemisphere.

RNA/Neurological Diversity, an epigenetic phenomenon that creates an infinitely diverse range of mindsets through the neurological gender synthesis (NGS), of the two hemispheres of the brain. This creates in each person a unique neurological gender ratio (NGR) (not heritable).

Acting in complementary fashion, these two dimensions result in human resilience at the scale of the individual (DNA/genetic) and the species (RNA/epigenetic).

GENDER

Gender is a human biological characteristic that is expressed in two forms: as anatomical gender (male or female) of the body, and as neurological gender (maleness *and* femaleness) of the brain.

FREE WILL

The self-aware mind assesses ongoing survival/opportunity in the surrounding environment. The resulting phenomenon is experienced as a three-way internal "conversation" among the two ancestral genders of maleness and femaleness (the hemispheres of the brain) that constitute *Nature*, and the contemporaneous voice of our own free will, which constitutes *Nurture*. This is a deliberative phenomenon.

Figure 6.3. Cave Art—Last Days of the European Hunter-Gatherer (6,000 YBP)

Note: Elongated Ethiopian body type and long bow.

Observations in Passing

AS MENTIONED IN THE FOREWORD, THE OBSERVATION OF EVENTS in my day-to-day life was the motivation to write *Two Minds*. We all have very different pathways through life and we all view those experiences through our mind's unique prism of gender. Nevertheless, there are lessons to be learned in each person's formative events, and therefore, I share these retellings from my own path.

OBSERVATIONS ON WAR

I was born on July 21, 1944, while Operation Neptune (The Allied Invasion of France: D-Day) was nearing completion. This was, of course, World War II, the war immediately after "The War to End all Wars," World War I. My first personal recollection of a war was the Korean War which started in 1950, the summer before I went to the 1st grade. The cease-fire came in 1953 (my 3rd grade). The Vietnam War began in the fall of 1959 when I was a sophomore in high school and ended in 1975, long after I had finished college and moved to New York City. In the interim, came the Cuban Missile Crisis of 1962 and then there was the Arab-Israeli War (the "Six Day War") in 1967. I well remember the Soviet Invasion of Afghanistan in 1979 and our backing of the Mujahideen. This was one of the final examples of the Cold War enmity and its related nuclear threat that loomed over the childhood and early adulthood of anyone my age in the U.S.A. The Soviets withdrew from Afghanistan in 1989, the Persian Gulf War followed shortly in 1991, and the collapse of the USSR occurred at the end of 1991.

At 10:28 AM on the morning of September 11, 2001 standing with a deeply shaken group of on-lookers in the roadway of Fifth Avenue in front of the New

York Public Library, I witnessed the collapse of the North Tower of the World Trade Center. For over two days in the chaos that followed, I was unable to determine if friends and colleagues who worked at the North Tower had made it out alive—they had. Shortly thereafter we invaded Afghanistan (October 2001), achieved a temporary unseating of the Taliban and, in March of 2003, initiated war against Iraq. Both conflicts continue at some level as of this writing.

I do not recount this chronology of nearly continuous war to suggest that we are a uniquely war-like nation or bent on world domination. Rather, it is a tapestry of competition that common sense calls us to recognize as a native feature of *Homo sapiens* striving. Since my reporting above is of the most modern, civilized and humane period of *Homo sapiens* on earth, it is not difficult to imagine how such striving was conducted in the Paleolithic Era, absent national borders and laws of conduct.

OBSERVATIONS ON DISCRIMINATION AND RACISM

In the summer following first grade in Lincolnton, North Carolina, my younger sister and I went to live with our aunt and uncle in Richmond, Virginia (our father was recovering from a stroke). One of our first experiences, shortly after our arrival, was attending a summer day camp—Camp Sioux. Looking back, I feel confident that the Camp was not named for the prowess of the Sioux at The Battle of the Little Bighorn (Custer's Last Stand), but rather the resourcefulness of Native Americans in natural settings, lessons for this band of kids. A number of large mounds were in the forest bordering our open play area (they were, as it later turned out, accurately proclaimed to be Native American burial mounds). One of these became the scene of an early but very memorable "us" versus "them" event.

Day One found me trying to mix in with my fellow all-white eight-year-olds who immediately recognized me as being from "out-of-town."

"Where are you from?"
"I am from Lincolnton, North Carolina."
"**North** Carolina!!!!... You're a Yankee!"

Despite my general protests (being equally uniformed on the geography of the matter), I was assigned to the "army" that would start from the foot of the mound (disadvantaged) to try to unseat the "army" at the top of the mound (advantaged).

You almost have to laugh at the layers of bias and irony in this small event. This minor dust-up occurred among a cluster of 7–9 year-old white males sitting in the 1950's lap of the segregated South, running under the banner of the Sioux Nation, discriminating among themselves based on imagined connections to a

Civil War (one hundred years past), to decide who was to get the lesser starting point in a make-believe war conducted on a Native American Burial Mound.

After returning from Virginia, my sister and I grew up in Lincolnton, North Carolina (pop. 5,000), a typical segregated Southern town of the 50's and early 60's. While I did not consider my parents, or particularly the matriarch of the family, my grandmother, to be racist, over time I began to feel that even their good works and good intentions were a form of condescension to blacks. In my mind, they just weren't embracing the vision of equality being articulated by Martin Luther King.

By the time I was mid-way through high school, the early sixties, the South was entering a period of racial strife and violence, with Birmingham, Montgomery, Selma, and much of Mississippi at center stage. Lincolnton, all the way up to my high school graduation in 1962, always seemed to be relatively calm as this storm raged. After I left to attend college, my growing commitment to civil rights became a point of friction with my family, and over time, a conviction grew on my part that the South was heavily, if not exclusively, racist.

In the summer of 1965 after my third year in college at North Carolina State, a fraternity brother and his best buddy from high school (both from New York) set out to travel around Western Europe. Our trip followed two years of increasing violence in the South and beyond:

Aug 1963   March on Washington: Dr. King's speech: "I have a dream"

Sept 1963   Bombing of Baptist Church by Ku Klux Klan Birmingham, Alabama. Four black girls killed in Sunday school.

Nov 1963   President John F. Kennedy assassinated: Dallas, Texas

June 1964   Three Civil Rights workers murdered by Ku Klux Klan (ages 20, 21, 24) supporting voter rights, in Meridian, Mississippi.

July 1964   Passage of the Civil Rights Act of 1964.

Mar 1965   Bloody Sunday" Selma, Alabama: 600 marchers were attacked, beaten and tear-gassed by State and Local lawmen, "…represented the political and emotional peak of the modern civil rights movement."

We arrived in France two months after "Bloody Sunday." Nonetheless, I wasn't prepared for the level of accusatory questioning that followed identifying myself as an American from the South.

One night from that summer stands out. We were at a restaurant in Bern,

Switzerland and the proprietor, Mr. H, sat with the three of us after closing while the staff cleaned up. Mr. H: "I still don't understand how this racist treatment can be based on the color of a person's skin." This comment followed his reference to a number of affluent, college-educated young blacks and a Jazz musician that he knew in Bern. I replied that the situation was much more complicated and that after two hundred years of slavery and one hundred years of discrimination, Southern blacks had a very narrow choice of jobs, usually manual labor, and, as in any group of people so marginalized and economically disadvantaged, they had a higher frequency of alcoholism, divorce and illiteracy. Many Southern whites said this behavior confirmed the inferiority of blacks rather than admitting that their own multi-generational racism was the root cause of this pervasive socioeconomic inequality.

Mr. H: "Sounds like the Northern Italians."

RRC: "What do you mean?"

Mr. H: "Well, we import Northern Italians to work on our roads and dams and they live in these work camps… drunks, violent, dirty…they're a danger to our community."

RRC: "Not all Northern Italians are like that, are they?"

Mr. H: "They're animals!"

No amount of pointing out Mr. H's discrimination and stereotyping of Northern Italians as another version Southern racism could get him to budge an inch. His position was rational while the American South was racist!

This and similar encounters quickly changed my expectation of tolerant Europeans vs. the ugly Americans.

Another lesson on the near-universality of racial discrimination came in a taxi cab on the way into Logan Airport for my first job out of college in the summer of 1968, (finally making my escape from the "racist-plagued" South). That summer later became known as the "Year of the Pitcher" and Bob Gibson, one of the greatest black athletes and pitchers of all time (1967 Cy Young Award; 1968 Most Valuable Player Award National League; 1968 World Series Record for Strike-outs) was tearing up baseball's record books.

As we headed into Boston, my cab driver launched into his observations on the baseball season and quickly moved on to the subject of Bob Gibson… out of his mouth came an endless stream of racist hatred and epithets as crude and shocking as anything I had encountered in the South—all leading to the conclusion that Gibson had ruined the game of baseball. This outburst was the beginning of the end of my myth of northern racial tolerance.

And as for that illusion of my hometown as a relative island of calm in the midst of racial strife? That came apart on a morning in November 1979 as I read *The New York Times* front page…five marchers shot and killed, 10 wounded in a mid-day attack by members of the Ku Klux Klan in Greensboro, North Carolina! As I read the names and home towns of the perpetrators, "Lincolnton" jumped off the page for several of them.

It got worse, in the detailed follow-up by PBS's *Frontline* on January 24, 1983, it was confirmed that Virgil Griffin, Grand Dragon of the Ku Klux Klan, and his men had attended a key rally and recruitment meeting to organize the Klan's bloody response to the Greensboro march…that meeing was held in Lincolnton, North Carolina.

As a final incident under the "discrimination and racism" banner, I would like to return to the white-on-white discrimination witnessed at Camp Sioux. In this case, however, it is female-on-female. Phoebe Prince was a lovely fifteen-year-old recent Irish immigrant attending South Hadley High School in Massachusetts. In March of 2010, she came home from school and hanged herself. Six teens, mostly teenage girls, were prosecuted in her death, having been accused of instigating or participating in months of torment of Phoebe—"merciless and sometimes violent bullying."

In addition to being from out-of-town and out-of-country and out-of-culture, Phoebe had apparently dated one of the football players at the High School—setting off a no-holds-barred competition with the local clique of teenage girls. Aside from the extreme outcome (the suicide of this young girl), this is a common type of social and sexual competition among teenage girls.

As we look back to the playgrounds of our youth, to the recess periods and our passive participation in the behavior of the group, many memories of cruel treatment still linger. These were actions that perhaps we didn't start, but we did nothing to stop. I remember our playground pack of second grade boys at Ginter Park Elementary School in Richmond, Virginia, running from and isolating a kid with a paralyzed left arm, and the fourth grade version of the same discrimination at Lincolnton Grammar School in North Carolina; naming the boy with the colostomy bag "Jimmy Hospital"—excluding him in a similarly cruel fashion.

There is a deep-seated urge to compete, to seek revenge for perceived insult, or even to seek dominance and status among peers for its own sake which is a persistent streak in *Homo sapiens* and certainly not just an isolated failure in South Hadley, Massachusetts. A more accurate description of this tragic event is that it represents characteristic *Homo sapiens* behavior that was allowed to expand beyond the limits that we have imposed and enforce as a civilized society.

A couple of observations on the inherent nature of competition, discrimination and racism:

> *Homo sapiens*—males and females who are *maleness* dominant—will, even at an early age, strive for social status and for the lead role among peers. Competitive discrimination aimed at fellow humans, not considered to be peers, may be physical appearance (even exaggeration of minor differences), nationality, gender, race, sexual preference, age, clothing, or any combination of the above, or an endless list of other **real or imagined or fabricated** aspects.

Competition is an over-arching feature of *Homo sapiens* life, and is an inherited evolutionary imperative. Civilizations counter-balance the excesses of competition through the articulation of humanistic ideals, laws, and the enforcement of human rights.

OBSERVATIONS ON GENDER

In December of 1968, I arrived in New York City to begin work with the architectural firm of I. M. Pei & Partners. As a member of the Post WWII Baby Boom, I arrived with the full set of my generation's anti-establishment if not revolutionary values and beliefs—a major irritant to the 1950's social and cultural tranquility of America. Many changes were already underway, including Civil Rights, the Anti-War Movement and the Sexual Revolution, all of which seemed to launch just as we were growing out of the awkward phase of puberty, and riding the crest of the golden era of Rock and Roll. The fact that the birth control pill was added to the mix (first introduced and approved in 1960) did little to calm things down.

Another challenge to the status quo was the movement for Women's Rights which had been framed in no uncertain terms by Betty Friedan's "The Female Mystique" in 1963. Shirley Chisholm began her service as the first black woman elected by Congress in 1968 and she advocated for the Equal Rights Amendment to the Constitution, passed by Congress in 1969 (although not ratified by the States). Gloria Steinem launched *MS Magazine* in 1972 and the Supreme Court's decision on Roe vs. Wade legalized abortion in America 1973.

This tumultuous period in American history forced humanistic values of the change-the-world type into the face of the status quo, however, as in all revolutions (even our mini revolution), there was a "little red book" of received wisdom that comes with membership. One of the commandments was that men and women were equal, not only equal in rights, but equal in every way. If parents and society didn't project those male and female roles on children from birth—giving little boys guns and little girls' dolls—they wouldn't grow up into male and female stereotypes. Yes, I said it. I was 100% on board and my wife, for example, had a strong career as a financial advisor and served on the founding board of the First Women's Bank of New York (1975). All aspects of life seemed to fit the picture of this brave new world.

And then, a little over eighteen years after arriving in Manhattan, I confronted nature's great awakening moment…babies! My two sons and their adventures growing up completely changed my view about the possibility of profound biological differences among boys and girls; two of those moments standout over all others. At the tender age of two years, two months, my son William was being stuffed by me into a seersucker dress suit (summer shorts and white socks) for, of all things, a competitive interview for school admission! For all people in the normal world, I need to explain that in order to get into a Pre-K (Pre-Kindergarten) program in Manhattan; one enters into a general feeding frenzy for the small number of openings.

One of the early hints about gender differences was that girls typically potty-trained earlier than boys. This interview was coming at that last moment of uncertainty where William had pretty much mastered the potty routine, but I was nervous. With visions of the interview from Hell hanging over me, I decided that the prudent thing was to slip a diaper under those summer shorts, just for safety's sake. Although, let's face it, no matter how confident he is in the interview, no matter how much he impresses the teacher, wearing a diaper undercuts the "Master of the Universe" message!

Nonetheless, we entered the hallowed halls of Brick Church School on time and, since on any given day William's behavior could fall anywhere between Mother Teresa and the Exorcist, I was relieved that he seemed to be somewhere in the middle.

As we entered the large sunlit classroom, I noticed that (due to the crush of interviews) there were two children-height tables, each with those impossibly small chairs, set up about fifteen feet apart so that two teachers could conduct simultaneous interviews. We entered the room on the left, and as the very correctly attired teacher stood up and extended her hand to shake mine, I noticed an elaborate multi-story toy zoo, a class project, to the left. I suddenly felt as though my left hand and arm were being jerked out of my shoulder socket…I had JAWS on the end of my arm! I looked down at William (at this point a wild animal straining at the leash) and then looked up at the object of his desire…delicately stacked building blocks forming five levels of cages with realistic brightly colored circus animals in each of dozens of cages. Of particular interest to me were the large yellow signs which cried out in red Pentel…DO NOT TOUCH!!! CLASS PROJECT!! DELICATE!!

Somehow (my memory of this entire event goes in and out on me) I managed to get William back to the table and in a seat—without using the threat of bodily harm in front of the teacher. And then it happened, the event that communicated to me once and for all that when anatomical and neurological gender converge (a high *maleness* boy and a high *femaleness* girl), you are in for vast differences of perception and behavior.

The teacher to our right was rising to greet a parent (a mother) who was entering the classroom with her 2+ year-old daughter. Right away I knew something was off, *the mother was not holding the child by the hand…the daughter was loose in the room!* Calmly, her daughter walked over to the table while her mom was speaking with the teacher.

That's when I saw something I will never forget. The little girl pulled back a chair, took off her white cardigan sweater, folded it lengthwise, placed it over the back of the chair and sat down to wait for her mom. A thousand questions were running through my mind at that moment…Does she use whips? Are drugs involved (and where do I get them)? Does she have a cattle prod at home? And then it hit me, all those hints at the playground in Central Park, the birthday party for the little girl in 6A…No! There was no whip, no drugs, no electric shock. These girls and boys were just profoundly different from one another, a difference that nurture cannot overturn (Note: one positive footnote: William was accepted…there is apparently quite a wide threshold of acceptable behavior for little boys!).

Another example of the avalanche of such moments occurred with my younger son, Beau, in Sag Harbor, NY on a Saturday morning while leaving a small neighborhood store, Espresso. Beau was about two years old when, as I pushed his stroller down the side street, something caught his eye in the yard to the left. The owner had an apartment behind the store.

As we came to a break in the hedge there it was, a gleaming black and stainless steel Harley-Davidson motorcycle that someone had just finished washing… water still dripping off in the sun. Well, when Beau wanted to immobilize the stroller so I couldn't move at all, he would go stiff as a board with his legs extended out over the front of the stroller, braced on the ground. Beau was fixated on that motorcycle, which was, of course, sitting on private property, while I was determined to get back to the house. We were already running late so Beau, sensing that, had immediately assumed the position!

As I stopped and reached down to try to get his extended legs back in the stroller, an unforgettable event occurred. Beau somehow got to his feet and was now standing, and wearing the stroller! I reached out to pull the stroller back down to the ground and then thought better of it, realizing that I was witnessing an Olympic Moment! Beau, in the manner of Frankenstein's monster was, step-by-halting-step, walking into that yard and toward that motorcycle, arms outstretched with grasping fingers extended. The handle bars swayed widely left and right in the air as I paused in fatherly pride to see this latest attraction to another massive, shiny macho object that I was confident his little boy mind (high *maleness*), could not possibly understand.

So, as happens with parents all over the world, my education on the true nature of a child learning and growing up (which is so invisible when you are the child who is doing it) proceeded over the years until my graduation with a full appreciation of how inaccurate my naïve preconception of biological uniformity of men and women had been in 1968. Of course, my larger ongoing education was flowing from the pursuit of a professional design career in New York City and working with a vast array of talented people from many different nations, races, and all variations in the range of neurological genders I describe in *Two Minds*.

Looking back over my life, I realize that I also experienced an exemplar of the *maleness–femaleness* phenomenon in the two towering figures of the Civil Rights Movement: Martin Luther King, Jr., who defined the non-violent and New Testament strategy, displaying equal measures of empathy, restraint and oratorical power, while Malcolm X, out of the "righteous vengeance" pages of the Old Testament, demanded in a militant manner, an eye-for-an-eye! Both in the gender realities of their era, were anatomical males, yet we find in them the deep-seated and expressed differences in *maleness–femaleness* and associated traits reflected in us all.

Malcolm X provided one of the most dramatic examples of the exercise of free will, going against a lifetime of aggressive and isolationist rhetoric. Leading up to his assassination, following a pilgrimage to Mecca, he underwent a conversion from his original Black Nationalist views...his last words supported a brotherhood of man and he spoke his truth in the face of the violent Muslim Brotherhood... the initiating event of his assassination.

The conclusion I have drawn and have sought to more fully understand in the writing of *Two Minds*, is that we currently have a misconception of gender, and have failed to understand its foundational role in human origins, the survival of our species, and the formation of our human natures. While I have articulated a much more complex role and integrative effect across the human population for gender, I am fully aware that this too is simplification, a working approximation on the way to a more complex and fully developed understanding. As Iain McGilchrist wrote when describing the world as perceived by the right hemisphere of the brain, *"Truth is provisional."*

Acknowledgements

Above all great works and all great minds, foundational and inspirational to the making of this book, I acknowledge Edward O. Wilson, University Professor Emeritus, Harvard University. A gracious first critique of the 2012 manuscript by Donald Lamm, past chairman of W. W. Norton & Company, was instrumental in the distillation of this work and the confidence to find my own voice. Janet Adams Strong, a fearless editor of wayward writers, guided my wanderings into the more straightforward path of its final form. Don Watson and Robert Koester served the invaluable and counter punctual role of sage adviser/critics, and, of course, I must thank Dr. Carl McDaniel, whose rigorous critique prompted a deeper reconsideration and further advancement of the work, and Dr. Miriam Feuerman who asked *the question*.

Randolph R. Croxton
New York, New York
June 2015

List of Figures

List of Tables

Notes

INTRODUCTION

1. International Human Genome Sequencing Consortium, "Initial Sequencing and Analysis of the Human Genome," *Nature* 409 (February 15, 2001), 860–921.
2. Edward O. Wilson, *The Social Conquest of Earth* (New York: W. W. Norton, 2012), 89.
3. Michael Snyder, "Research Highlights, Genome Characterization and Gene Regulation," *Michael Snyder Lab, Number 8*: "…regulatory sequences are quite divergent between species and between individuals—much more than protein coding sequences."
   Updated by Xiyan Li (650-723-9914) July 7, 2014; accessed July 22, 2014, http://snyderlab.stanford.edu/resume10.html.
4. J. Kocerha et. al., "Noncoding RNAs and neurobehavioral mechanisms in psychiatric disease," *Molecular Psychiatry* (March 31, 2015)/doi:10.1038/mp.2015.30
   Quote from Abstract:
   "It is now abundantly clear that neurobehavioral phenotypes (individual human behavior) are epigenetically controlled by non-coding RNAS (ncRNAs). The micro RNA (miRNA) class of ncRNAs is ubiquitously expressed throughout the brain and governs all major neuronal pathways."
5. Kelly, Kevin. *Out of Control: The New Biology of Machines, Social Systems and the Economic World.* (Boston: Addison-Wesley, 1994), 98.

CHAPTER ONE

1. Per Sjödin et al, "Re-sequencing Data Provide No Evidence for a Human Bottleneck in Africa during the Penultimate Glacial Period," *Molecular Biology and Evolution* 29 no. 7 (July 1, 2012), 1851–60. [Author's note: This paper confirms population crash approximately 130,000 YBP in association with MIS6 climatic variances, drought, and volcanic winter, and also the resulting significant reduction in genetic diversity, but argues against a genetic "bottleneck" per se.]
2. Paola Villa et al, "The Still Bay Points of Blombos Cave" (South Africa). *Journal of Archaeological Science* 36 (2009), 441-460. Cf. also Marlize Lombard and Laurel Phillipson, "Indications of bow and stone-tipped arrow use 64,000 years ago in Kwazulu-Natal, South Africa," *Quarterly Review of World Archaeology* 84 no. 325 (2010): 635–648 and C.S. Henshilwood, "A 100,000 year old ochre-producing workshop at Blombos Cave, South Africa," *Science* 334 no. 6053 (October 2011): 219–222, DOI: 10.1126/Science.1211535.

3.   Matthew L. Sisk and John J. Shea, "The African Origin of Complex Projectile Technology: an Analysis Using Tip Cross-sectional Area and Perimeter," *International Journal of Evolution Biology* 2011 (Epub March 30, 2011): Article ID 968012.

4.   Kate Wong, "Human Evolution: Inside the Neanderthal Mind," *Scientific American* (February 2015), 36–43.

5.   Liane Gabora and Kristy Kitto, "Concept Combination and the Origins of Complex Cognition," *Origins of Mind: Biosemiotics Series*, vol. 8, edited by Liz Swan (Berlin: Springer, 2013): 361–382. [Author's Note: A link between "seeing in context" and self- awareness].

6.   Chris Stringer, Lone Survivors: *How We Came to be the Only Humans on Earth* (New York: Time Books, Henry Holt & Company, 2012), 207.

7.   Johannes Krause et al, "The Complete Mitochondrial DNA Genome of an Unknown Hominin from Southern Siberia," *Nature* 464 (April 8, 2010): 894–897, DOI: 10.1038/nature08976.

8.   Richard Green et al, "A Draft Sequence of the Neanderthal Genome," *Science* 328 no. 5979 (May 7, 2010): 710–722, DOI: 10.1126/Science 1188021.

9.   Rebecca L. Cann, Mark Stoneking, and Allan C. Wilson, "Mitochondrial DNA and Human Evolution," *Nature* 325 (January 1, 1987): 31–36, DOI: 10.1038/325031a0.

10. Clive Finlayson, *The Humans Who Went Extinct: Why Neanderthals Died Out and We Survived* (Oxford: Oxford University Press, 2009).

11. Brian Fagan, *Cro-Magnon: How the Ice Age Gave Birth to the First Modern Humans* (London: Bloomsbury Press, 2010).

12. Richard Potts and Chris Sloan, *What Does It Means to be Human?* (Washington, DC: National Geographic, 2010).

13. Jared Diamond, *The Third Chimpanzee* (New York: Harper Collin, 1992); *Guns, Germs and Steel* (New York: W.W. Norton, 1997); *Collapse* (New York: The Viking Press, 2005).

14. Stephen Pinker, *The Better Angels of our Nature: Why Violence has Declined* (New York: Viking, 2011).

15. Wilson, *The Social Conquest of Earth*, op. cit.

16. Stringer, *Lone Survivors*, op. cit.

17. Ruth C. Engs, *The Eugenics Movement: An Encyclopedia* (New York: Greenwood Press, 2005)

18. Edward O. Wilson, *Sociobiology: The New Synthesis* (Cambridge, MA: Harvard University Press, 1975)

19. B. F. Skinner, *About Behaviorism* (New York: Alfred A. Knopf, 1974), 37–50.

20. Michael Balter, "Humans and Neandertals likely interbred in Middle East," *Science* DOI: 10.1126/science.AAA6410 (January 28, 2015).

21. Paul Mellars and Jennifer C. French, "Tenfold Population Increase in Western Europe at the Neanderthal-to-Modern-Transition" *Science* 333 no. 6042 (July 29, 2011): 623–627.

CHAPTER TWO

1. Lawrence H. Keeley, *War Before Civilization: The Myth of the Peaceful Savage*, Oxford: Oxford University Press, 1996).

2. Diamond, *Guns, Germs and Steel*, op.cit.

3. Iain McGilchrist, *The Master and His Emissary: The Divided Brain and the Making of the Western World* (New Haven: Yale University Press, 2009).

4. Keeley, *War Before Civilization,* op. cit., 38.

5. Stephen E. Churchill et al, "Shanidar 3 Neanderthal rib puncture wound and Paleolithic weaponry" *Journal of Human Evolution* 57 no. 2 (August 2009), 163–178.

6. Joseph G. Jorgensen, *Western Indians: Comparative Environments, Languages, and Cultures of 172 Western American Indian Tribes* (San Francisco: W. H. Freeman & Co, 1980).

7. Pinker, *The Better Angels of our Nature*, op. cit.

8. Keeley, *War Before Civilization*, op. cit., 68.

9. Philip L. Walker, "A Bioarchaeological Perspective on the History of Violence," *Annual Review of Anthropology* 30 (October, 2001), 590, doi: 10.1146/annurev. anthro.30.1.573.

10. Ibid., 573.

11. Keeley, *War Before Civilization*, op. cit., 69.

12. Diamond, *Guns, Germs and Steel*, op.cit., 16.

13. The hunter-gatherer culture of humankind extends through the Paleolithic Era and to the end of the very brief Mesolithic Era 11,500 to 7,500 YBP. For this reason, the final settlements of the hunter-gatherers are referred to as "Mesolithic."

14. Mark Golitko and Lawrence H. Keeley, "Beating Ploughshares Back into Swords: Warfare in the *Linearbandkeramik*" *Antiquity* 81 no. 312 (June 2007), 332–342, doi: http://dx.doi.org/10.1017/S0003598X00095211

15. Ibid., 332.

Peter Rowley-Conway, "Westward Ho! The Spread of Agriculture from Central Europe to the Atlantic," *Current Anthropology* 52 no. S4 (October 2011), S431–S451, www.jstor.org/stable/10.1086/658368.

16. Golitko and Keeley, "Beating Ploughshares Back into Swords," op. cit.

17. Rowley-Conway, "Westward Ho!," op. cit.

18. T. Zerjal et al, "The Genetic Legacy of the Mongols" *American Journal of Human Genetics* 72 no. 3 (March 2003), 717–721.

19. Michael Gibson, *Genghis Kahn & the Mongols* (New York: Sentinel Books, 1973).

20. Anne E. Pusey and Craig Packer, "Infanticide in Lions: Consequences and Counterstrategies," *Infanticide and Parental Care*, ed. Stefano Parmigiani & Frederick S. vom Saal (Chur, Switzerland: Harwood Academic Press, 1994), 277–299.

21. Julie A. Jimenez et al, "An Experimental Study of Inbreeding Depression in a Natural Habitat," *Science* 265 (October 14, 1994), 271–273.

22. Elia Roberts et al, "A Bruce Effect in Wild Geladas," *Science* 335 no. 6073, 1222–1225, February 23, 2012, doi:10.1126/science.1213600.

23. Debra Lieberman et al, "The Evolution of Human Incest Avoidance Mechanisms: An evolutionary psychological approach," *Center for Evolutionary Psychology, University of California, Santa Barbara* (October 2000), http://citeseerx.ist.psu.edu/viewdoc/download?doi=10.1.1.140.8222&rep=rep1&type=pdf.

24. "Twelve Tables of Roman Law," *Colliers Encyclopedia* (1921).

25. Saioa Lopez and Santos Alonzo, "Evolution of Skin Pigmentation in Humans" *Wiley Online Library/eLs* (June 16, 2014), doi: 10.1002/9780470015902. a0021001.pu b2.

26. David Reich et al, "Ancient Human Genomes Suggest Three Ancestral Populations for Present Day Europeans," *Nature* 513 no. 7518 (2014), September 18, 2014, 409, doi: 10.1038/nature13673.

27. David Reich et al, "The Genomic Landscape of Neanderthal Ancestry in Present-day Humans," *Nature* 507, 354–357 (2014), March 20, 2014, doi:10.1038/nature12961.

28. David Reich, Svante Pääbo et al, "The Complete Genome Sequence of a Neanderthal from the Altai Mountains," *Nature* 505 (2013), January 2, 2014, doi: 10.1038/nature12886.

29. Reich, "Ancient Human Genomes," op. cit.

30. McGilchrist, *The Master and His Emissary*, op.cit.

31. William E. Banks et al, "Neanderthal Extinction by Competitive Exclusion" PLOS ONE 3(12): e3972; doi:10.1371/journal.pone.0003972.

CHAPTER THREE

1. Stephen Prothero, *God is Not One: The Eight Rival Religions that Run the World* (New York: Harper Collins, 2010).

2. Lionel Tiger and Michael T. McGuire, *God's Brain* (Amherst, NY: Prometheus Books, 2010), 15.

3. Marie-Luise Mechias et al, "A Meta-analysis of Instructed Fear Studies: Implications for Conscious Appraisal of Threat," *NeuroImage* 49 (January 15, 2010), 1760–1768, doi: 10.1016/neuroimage.2009.09.040. Cf. also N. Kinnear, S.W. Kelly et al (2013), "Understanding how Drivers Learn to Anticipate Risk on the Road: A Laboratory Experiment of Affective Anticipation of Road Hazards," *Accident Analysis and Prevention* 50 (January, 2013), 1025–1033, doi: 10.1016/j. aap.2012.08.008. "A differentiation between cognitive and psycho-physiological responses was also found supporting theory that distinguishes between conscious and non-conscious risk appraisal."

4. Jarrett A. Lobell, "New Life for the Lion Man" *Archaeology* 65 no. 2 (March/April 2012); http://archive.archaeology.org/1203/features/stadelhole_hohlenstein_paleolithic_lowenmensch.html. Cf. also Martin Bailey, "Ice Age Man is World's earliest Figurative Sculpture," *The ARTNewspaper*, 31 (January 2013).

5. James Wasserman, *The Egyptian Book of the Dead: The Book of Going Forth by Day* (San Francisco: Chronicle Books, 1994).

6. *NASB Study Bible*, ed. Kenneth L. Barker et. al. (Grand Rapids, MI: Zondervan, 2000).

7. J. E. Taylor, "The New Atheists," *Internet Encyclopedia of Philosophy*, IEP (ISSN 2161-0002 (2010), http://www.iep.utm.edu/n-atheis/.

8. Hal Hellman, *Great Feuds in Science: Ten of the Liveliest Disputes Ever* (New York: John Wiley & Sons, Inc., 1998).

9. *Marx's Critique of Hegel's Philosophy of Right*, [1843], ed. Joseph O'Malley (Cambridge University Press, 1970), Introduction.

10. Valerie A. Andrushko et al, "Investigating a Child Sacrifice Event from the Inca Heartland," *Journal of Archaeological Science* 38 no. 2 (February 2011), 323–333.

11. Pinker, *The Better Angels of our Nature*, op. cit.

CHAPTER FOUR

1. Joe Clark, "Gay Money: The Truth about Lesbian and Gay Economics," November 2010, joeclark.org/gay money/findings/: compilation and interpretation of 70 research papers on lesbian and gay economics: "Full findings/choices of Occupation."

2. Paul E. Ehrlich, *Human Natures: Genes, Cultures, and the Human Prospect* (Washington, DC: Island Press, 2000), 196.

3. Ibid., 197.

4. Ibid.

5. Gary J. Gates, "How Many People are Lesbian, Gay, Bisexual, and Transgender?," The Williams Institute, University of California School of Law (April 2011), 3. Cf. also Simon LeVay, *Gay, Straight and the Reason Why* (Oxford: Oxford University Press, 2011).

6. Louann Brizendine, M.D., *The Female Brain*, (New York: Broadway Books, 2006), 161.

7. Louann Brizendine, M.D., *The Male Brain*, (New York: Broadway Books, 2010).

8. Ibid., 133.

9. Ibid.

10. Ibid., 134.

11. Ibid.

12. Alicia Garcia-Falgueras and Dick F. Swabb, "Sexual Hormones and The Brain: An Essential alliance for Sexual Identity and Sexual Orientation,"*Endocrine Development* 17 (2010), 22–35.

13. McGilchrist, *The Master and his Emissary*, op. cit., 13.

14. Ibid.

15. NGR: Neurological Gender Ratio: The proportional influence of left vs. right hemispheres of the brain (i.e. "maleness" experiences/left hemisphere and

"femaleness" experiences/right hemisphere) in establishing the overall perception of the outside world and thereby the personal attributes of the individual. A wide range of non-sexual inclinations, skills and values are associated with each hemisphere of the brain based on gender experiences.

16. Anomalies of the Right Hemisphere: Two attributes which are physically located in the right hemisphere but, under strong selection pressure, have come to be expressed predominately in the male are the sense of directionality and distance to the source of a sound (such as someone stepping on a twig) as well as 3-dimenstional orientation as may be required in land navigation/topography or parallel parking. One speculation is that in the earliest formation of the bi-hemispheric brain, as in birds, the female of the species would have been protecting the nest. With the emergence of primates' larger brains and longer maturation periods, the necessity of males to hold territory and protect the perimeter of this expanded "nest" may have led over millions of years to the usurpation of these attributes as predominately male in humans.

17. Ogi Ogas and Sai Gaddam, *A Billion Wicked Thoughts: What the World's Largest Experiment Reveals About Human Desire* (New York: Dutton, 2011).

18. Catherine Salmon and Donald Symons, *Warrior Lovers: Erotic Fiction, Evolution and Female Sexuality* (New Haven: Yale University Press, 2001).

19. Ogas and Gaddam, *A Billion Wicked Thoughts*, op. cit., 47.

20. Ibid., 33.

21. Martha Stout, *The Sociopath Next Door* (New York: Broadway Books, 2005).

CHAPTER FIVE

1. Clara Moskowitz, "Gravitational Waves from Big Bang Detected" (March 17, 2014), *Scientific American*, www.scientificamerican.com/article/gravity-waves-cmb-b-mode-polarization.

2. Petra Pavšič, "War Rape: A Planned and Targeted Policy in War," *Consultancy Africa Intelligence (CAI)* (September 17, 2012), www.consultancyafrica.com/index.php?option=com_content&view=article&id=1115:rape-a-planned-and-targeted-policy-in-war&catid=60:conflict-terrorism-discussion-papers&Itemid=265

3. Genocide Intervention Fund, "Darfur: Gendered Violence and Rape as a Weapon of Genocide," National Center on Domestic and Sexual Violence (2005), www.ncdsv.org/images/darfurgenderedviolencerapeweapon.pdf.

4. Global Peace Index (GPI): The Institute for Economics and Peace (2014), www.economicsandpeace.org/research/iep-indices-data/global-peace-index.

5. Aleksandr Solzhenitsyn, *The Gulag Archipelago, 1918-1956: An Experiment in Literary Investigation* (Boulder, Colorado: Westview Press, 1998).

6. Davis Grubb, *The Night of the Hunter* (1953), Disruptive Publishing Edition (2005).

7. Robert S. Norris, *Racing for the Bomb: General Leslie R. Groves, the Manhattan Project's indispensable Man* (Hanover, NH: Steerforth Press, 2002).

8. James Hansen et al, "Assessing 'Dangerous Climate Change': Required Reduction

of Carbon Emissions to Protect Young People, Future Generations and Nature," *PLOS ONE* 8 no. 12 (December 3, 2013), vol 8 no.12, www.columbia.edu/~jeh1/mailings/2013/20131202_PopularSciencePlosOneE.pdf.

CHAPTER SIX

1. Sriram Sankararaman et al., "The Genomic Landscape of Neanderthal Ancestry in Present-Day Humans," (January 2014), *Nature*, doi: 10.1038/nature 12961.

2. J. Kocerha et al., "Noncoding RNAs and Neurobehavioral Mechanisms in Psychiatric Disease," *Molecular Psychiatry*, (March 31, 2015), doi:10.1038/mp.2015.30. Quote from Abstract: "It is now abundantly clear that neurobehavioral phenotypes (individual human behavior) are epigenetically controlled by non-coding RNAS (ncRNAs). The micro RNA (miRNA) class of ncRNAs is ubiquitously expressed throughout the brain and governs all major neuronal pathways."

3. Daniel M. Messerschmidt et al., "DNA methalation dynamics during epigenetic reprogramming in the germline and preimplantation embryos," *Genes and Development*, (2014) 28:812–28. (Cold Spring Harbor, NY: Cold Spring Harbor Laboratory Press).

4. M. M. Patten, et al., "The Evolution of Genomic Imprinting: Theories, Predictions, and Empirical Tests," published online in *Heredity*, (2014) 113, 119–28, doi: 10.1038/hdy. 2014.29; published online April 23, 2014
    a. *The Kinship Theory* proposes that imprinting is a result of conflicting objectives of the parents—wanting to achieve the highest evolutionary fitness for their genes (paternal antagonism).
    b. *The Sexual Antagonism Theory* proposes that "the parent of the same sex as the offspring will, on average, provide fitter alleles (inheritance) to that offspring compared with the other parents."
    c. *The Coadaptation Theory* proposes that a high level of fitness results because parents of the same sex as the offspring have had past selection events creating a "degree" of coadaptation with their offspring.

    Note: All three theories are attempting to define the "fitness" of their approach in order to conform to the one-dimensional framework of natural selection. To the contrary, genomic imprinting is a two-dimensional phenomenon of DNA (natural selection) and ncRNA (neurological gender diversity), simultaneously in effect.

5. Ian Cowell, "Epigenetics—It's Not Just Genes That Make Us," (2013), British Society of Cell Biology, www.bscb.org/learning-resources/softcell-e-learning/epigenetics-its-no-just-genes-that-make-us.

6. Hansen, "Assessing 'Dangerous Climate Change': Required Reduction of Carbon Emissions to Protect Young People, Future Generations, and Nature," op. cit.

www.ingramcontent.com/pod-product-compliance
Lightning Source LLC
Chambersburg PA
CBHW060318310326
41914CB00102B/1992/J